Upheaval!

Why Catastrophic Earthquakes Will Soon Strike the United States

JOHN L. CASEY
WITH DR. DONG CHOI
AND DR. FUMIO TSUNODA, DR. OLE HUMLUM

Upheaval!

Why Catastrophic Earthquakes Will Soon Strike the United States

Order this book online at www.trafford.com
or email orders@trafford.com

Most Trafford titles are also available at major online book retailers.

Cover photo by: NOAA/NGDC, J. Dewey, U.S. Geological Survey.

Print information available on the last page.

ISBN: 978-1-4907-7903-4 (sc)
978-1-4907-7904-1 (e)

Library of Congress Control Number: 2016919743

Trafford rev. 12/09/2016

Trafford PUBLISHING® **www.trafford.com**
North America & international
toll-free: 1 888 232 4444 (USA & Canada)
fax: 812 355 4082

Dedication

This book is dedicated to the team of geologists and researchers who asked me to form the International Earthquake and Volcano Prediction Center (IEVPC). They are among the most courageous people I know, and they are dedicated to saving lives by challenging the barriers of the scientific establishment through innovative earthquake and volcano prediction research.

They are

Dr. Dong Choi, Dr. Fumio Tsunoda, Dr. Ole Humlum, Dr. Masashi Hayakawa, Dr. Zanghao Shou
Dr. Giovanni Gregori, Dr. Arun Bapat, Dr. Natarajan Venkatanathan,
Dr. Valentino Straser, Dr. Lev Maslov, Bruce Leybourne

Contents

Foreword

Since 2008, I have had the opportunity to follow the quest of one man, John Casey, and his singular, yet intensely dedicated efforts against overwhelming odds to tell both the American people and the rest of the world to get ready for a potentially dangerous new cold climate. Recently, he has added to the body of not just climatology but also of geology with research that connects what is happening on the Sun, with catastrophic earthquakes and volcanoes here on Earth.

It has been my distinct pleasure to be a firsthand witness and chronicler to the enduring struggle of John and the scientists around the world who have supported him to get the real story published about what is happening with the climate. He has been under assault from the U.S. government, environmental extremists, a hostile and often slanderous mainstream media intent on making sure no one hears his important message. He has further been assailed by a corrupt climate science establishment that has become enslaved to the almighty dollar and the fame of being politically correct.

The efforts to stifle John, his research, and his Space and Science Research Corporation (SSRC), were, in the end, for naught. The thoroughness and idealistic pursuit of truth in climate change research led him inexorably to the development of the relational cycle theory (RC theory) of climate change, and later to write his first climate book, *Cold Sun*. The book was later acquired by Humanix Books, a subsidiary of Newsmax Media, Inc., who restructured and updated that book and published it as *Dark Winter* in 2014. By midsummer 2015, it was the number one bestselling climate book in the United States at Amazon.com. I am delighted to say this book was not about global warming; rather, global cooling!

The RC theory, according to my own online review, is without question the best climate change theory today, far surpassing the utterly failed greenhouse gas theory. How do I know this? The answer is simple. Since he began to make public highly controversial predictions on our climate in spring of 2007, I can confirm that he has never erred! His long list of important climate variation predictions have been spot on, and I have documented them with my online postings over the years for all to read. No other climate researcher has such a public track record of accurately predicting the Earth's climate. No one!

This includes all the so-called best scientists at NASA, NOAA, the EPA, and of course at the United Nations Intergovernmental Panel on Climate Change (UNIPCC).Though he was deserted by so-called friends after he predicted the end of global warming back in 2007–2008, he now has perhaps tens of thousands of loyal supporters and followers and a host of lasting friends and supportive distinguished scientists around the world.

Bottom line, through it all since 2007, he and his relational cycle theory of climate change have weathered the storm and come out on top.

John has also drawn the support from many honest scientists around the world who have come to his aid and joined with him in the past nine years. Their support to John attests to the purity of his research and his obvious ability to understand complex engineering or scientific problems.

As Italian theoretical physicist Dr. Giovanni Gregori has said, "John Casey approaches problems like a true scientist, who follows Leonardo Da Vinci, and he also knows how to explain concepts in a form that anyone can understand."

A highly respected Japanese professor emeritus of geology, Dr. Fumio Tsunoda, a coauthor of this book, has given one of the most glowing opinions of John's climate understanding in his book *Dark Winter* with the following: "The air has filled with lectures and rumors that our Earth is getting warmer. The author of *Dark Winter*, John Casey, has found evidence to the contrary. His work is quite a revelation that marks a step toward a new scientific civilization. This book adds a brilliant page to the history of science."

And from an Indian physicist, Dr. Natarjan Venkatanathan, who said, "*Dark Winter* is simply a great work! It throws new light into the climate patterns of the Earth. John's concepts will help people better understand nature and the full story of what is behind our climate changes."

Scientific endorsements just don't get any better than that.

The SSRC closed in August 2015 after completing its years-long mission of alerting the American people, the media, the science establishment, and the U.S. government of the need to prepare for the potentially dangerous new cold climate. No one is more referenced on the web with regard to the coming cold climate than John Casey. Under his new one-man consulting company, Veritence Corporation, he is expanding his technical analysis and commentary into other areas of science—the space program for example.

Still, the impending calamities of this cold epoch have caused him to intensify his work with a new focus on the study of the imminent threat we are all under from predicted catastrophic earthquakes.

That new area of geological concentration actually began some years ago. His track record of success in climate science research led to him being asked by a group of international geologists in 2011 to form a new company. In February 2012, the International Earthquake and Volcano Prediction Center (IEVPC) was founded with the mission of saving lives by improving the world's ability in predicting major earthquakes of magnitude 6.0 or larger, yet with the added qualifier of doing so up to months in advance of the actual occurrence of the quakes. Dr. Dong Choi, in Canberra, Australia, became its director of research and is coauthor of this historic book.

Once more John and with his new team at the IEVPC have been challenging the science establishment and the U.S. government by actually doing earthquake prediction. Though they were told they could not predict earthquakes, through successive test programs since 2012, they have indeed shown they can materially improve the probabilities of major earthquake prediction. Like his proven track record in climate prediction, now he and his IEVPC team have actually predicted major earthquakes well in advance, specifically calling out the correct epicenter, along with the correct depth, magnitude, and timing of these big temblors.

Those of you who have always felt the U.S. government was politically biased, myopic, insensitive to the needs of the people, or unable to see the next step forward in technological advancements can add the area of earthquake prediction to the long list of matters where our leaders have failed us.

It is, therefore, with great respect and appreciation that I have accepted John's request to write this foreword for his new book *Upheaval!* Just as with *Dark Winter*, he and his coauthors have created a book with a critically important message that every American and every citizen of planet Earth should read and heed.

If the geophysical predictions of John and his associates come to pass with the accuracy of John's past climate predictions, then everyone should begin at once to prepare for the coming catastrophic earthquakes.

If these predictions come true during the time frame and locations he has specified, the damage to some regions of the United States could reach biblical scale.

So too will the proven ongoing ability of John to understand the connection between cycles of the Sun and the Earth and thus continue his unbroken track record of accurately predicting the future.

Dr. Rich Swier
www.drrichswier.com

Preface

In 2011, I contacted John L. Casey via email and asked him to create a company made up of some of the world's best earthquake-prediction experts. His individual efforts to discern and publicize the truth about the real causes and effects of climate change based on solar activity had become internationally known and thus led me to him along with his own highly credible research that showed a strong relationship between earthquakes, volcanic activity, and the Sun. I have been the IEVPC director of research and cofounder since the IEVPC came into being February 2012.

His first book, *Cold Sun*, was unique in that it clearly made the connection between cycles of the Sun and the most destructive earthquakes and volcanic eruptions. I and other international scientists were all too happy to endorse *Cold Sun* and its later revised publication, *Dark Winter*.

Fortunately, John agreed to my request to create the International Earthquake and Volcano Prediction Center (IEVPC) after reviewing the work of some of the geologists and seismologists that were leaders in new research about earthquake detection.

Under John's able leadership, this history-making new seismic research company, (IEVPC), has gone on to develop a workable process for synthesizing a variety of earthquake precursors into what we have proven is a successful model for predicting quakes at magnitude 6.0 or greater.

Throughout this period of time, it has been a genuine pleasure working with John Casey and drawing on his unique understanding of how the Earth's climate operates. Together, as discussed in this book, along with outstanding contributions from Dr. Fumio Tsunoda, Dr. Ole Humlum, and others, we have cooperated on research and associated science papers that have been published online for anyone interested in this subject without the cost normally required by professional journals. Our combined research, which John points out in *Upheaval!* has been highly revealing in terms of when the Sun goes into solar hibernation. We see our worst earthquakes and volcanic eruptions at these times. My own research shows this correlation is not only undeniable but goes back in time over thousands of years as well. The Sun, by some means yet to be identified, has an intimate linkage with major geophysical events.

John L. Casey

Demonstrating the tie between solar activity and earthquakes and volcanoes is a foundational new step forward. We both believe it will make many citizens of the Earth one day grateful for having been provided with much advance warning and preparation time for these catastrophic events.

In *Upheaval!*, we will attempt to pass on the best, most reliable information we can about the first of two major threats we will face in the future—that of deadly earthquakes that we are about to experience.

May you enjoy and benefit from our book and the science it offers.

Dr. Dong Choi
Director of Research
International Earthquake and Volcano Prediction Center (IEVPC)

Introduction

Despite the progress made by mankind over the millennia, life is still filled with wonders and unending lists of questions of how things are the way they are. From the smallest subatomic particle to the vastness of the universe, we are confronted with a host of challenges in expanding our grasp of the meaning, reasons for, and processes of creation.

If every one of the seven billion plus humans on this planet were given the freedom, quality of life, and resources to pursue a solution to some great puzzle of human existence, their efforts would likely yield seven billion plus more unanswered questions. Such is the history of scientific and human advancement.

Since 2007, my focus has been dedicated to telling my fellow citizens the truth about what was happening with the climate. Research in that telling year led me to later write two books, *Cold Sun* (2011) and its remake, *Dark Winter* (2014), both of which contain the following position:

The theory of man-made global warming and climate change based on human greenhouse gas emissions is the greatest international scientific fraud ever perpetrated on the world's citizens.

The reason for restating this conclusion again in this book is that it forms one of the important pillars of the subsequent research that links solar activity with the Earth's most destructive earthquakes and volcanic eruptions. One cannot continue to believe that man controls climate change and at the same time accept that the Sun and major earthquakes are interconnected as proposed herein.

The book Upheaval! *is about the threat of catastrophic earthquakes we about to endure that are produced during periods of global cooling, not global warming.*

Global cooling are words unspoken, taboo within the current leadership in Washington, even among some conservatives who are content to dismiss man-made global warming on economic grounds but lack the courage to go further and tell what is really happening to the climate—once more leaving the public in the dark.

The catastrophic earthquakes we are about to face occur when the Sun goes into decades-long cold phases called solar hibernations—also called grand minimums—within the solar physics community. This discovery adds to the bitter crop-destroying cold that arrives with these hibernations. In this text, we will look at the most immediate threat—catastrophic earthquakes.

Because of the research done in climate change, we can now predict with much greater accuracy the next period when the Sun cuts back on the energy by which it warms the Earth, thus bringing cold climates. We call these periods solar minimums. When they are especially long and deep, we call them grand solar minimums or, my preferred term, solar hibernations.

Since cold climates are strongly linked to catastrophic earthquakes as you will see in this book, once we have identified the time frame of the next cold epoch, the next solar hibernation, we will then know when our worst earthquakes are about to strike within a relatively narrow range of years. That is the central premise of this book.

The next solar minimum has started. It has begun with the end of our current 11-year solar cycle, number 24. The solar physics world began numbering the 11-year solar cycles starting in the mid-1700s. We are facing yet another solar hibernation of 20–30 years duration almost 200 years after the last such event. This solar minimum, a pronounced solar hibernation, has already been named the Eddy minimum by the solar physics community. The late Dr. John Eddy was a pioneer in the study of solar cycles.

The next two 11-year solar cycles, number 25 and 26, will bring the next deep cold epoch and with it the most dangerous period of earthquakes our kind has faced in over 200 years.

It is understandable there are many good Americans who may still remain shackled to decades of global warming propaganda, force-fed to us by self-serving politicians, corrupt scientists, and an obliging media. They have all misused a weak scientific theory, the greenhouse gas (GHG) theory, as a convenient tool to achieve their nefarious political objectives. For those Americans, there is this proven advice. Hedge your bets.

Take what you can accept from this book and others like it and make sure you are not caught unprepared on the wrong side of the Mother Nature's wrath. Feel free to skip most of the chapters of this book that may run afoul of your political sensitivities and go straight to the chapter titled "Preparing for the Great Upheaval." If you follow the recommendations there, you will be better able to withstand any future disaster, be it a financial, political, pathological, social, military, or geophysical event that may come along.

There are plenty of warnings of rough times ahead to be concerned and prepared for, coming from science, economic, medical, and military experts with causes completely unrelated to the Sun's control over the Earth. At the same time, one does not need to sign on to the host of unsubstantiated natural disasters, like rogue planets or "new world order" style of online conspiracies that are out there in order to recognize the future may not be so kind to us.

Thus, the chapter on preparedness may be lifesaving guidance without the need to change one's ideology or opinion of how climate variation occurs. You will be better prepared for the host of uncertainties that life is about to unleash upon you.

****

This book again, is about why, how, and when catastrophic earthquakes strike. And most importantly, can anything be done to provide substantial advance warning of these events so that we can be prepared, potentially saving many thousands of lives?

So how did this pursuit of geophysical threats begin?

In the midst of doing climate research back in late 2011, a group of seismology experts from the other side of the Earth asked for help in pulling them together into an earthquake-prediction research company. Their objective was to have an enduring effort toward solving the earthquake-prediction riddle along with the associated secondary question of when deadly volcanoes strike. That new organization, the International Earthquake and Volcano Prediction Center (IEVPC), went operational in February 2012. Its web site is at www.ievpc.org.

We launched our first test program for earthquake prediction right away with astounding success! This initial exposure to seismology led to an important conclusion. These talented experts, not individually but collectively, could in fact predict calamitous earthquakes to a degree never before achieved and do so with enough advance warning to save many lives.

Adding a climate connection and integrating the individual scientist's techniques into a comprehensive system for earthquake prediction then provided a more complete understanding for achieving the goals of the IEVPC. The resultant earthquake prediction methodology relies upon a multi-parameter system of earthquake-precursor signals.

It was great fortune in the process of setting up the IEVPC in 2012, and beginning its test programs, that support and advice was provided by my new friend and colleague Dr. Dong Choi, who is also coauthor of this book along with eminent geologist Dr. Fumio Tsunoda. Together we began a much more thorough effort to study the relationship between climate change and these geophysical events. It was quite unanticipated that a gold mine of new science that would result.

Here we are then, five years on into the study of the relationship between the Sun and the Earth's geological processes, though with much new documented research. The strong correlation between the Earth-Sun climate system and concurrent major earthquakes and volcanic eruptions as well now seems more valid and relevant than ever. Though the mechanisms for these geological threats are not well understood by the world geological community, we do at least know much about the timing of these events in relation to climate variation.

Much of this knowledge stems first from the ability to improve the prediction of climate change as has been shown using the relational cycle theory (RC theory). It is spelled out in appendix 1 at the back of this book. The RC theory in essence says that mankind has little to do with climate change and that the Sun drives the majority of climate variation.

So what will the reader find in this book? In Upheaval! we will generally explore four important themes:

1. That the cold climate phases of the Sun, called solar minimums or their more intense versions, solar hibernations, are coincident with the worst earthquakes. In a solar minimum, the Sun dramatically reduces the energy by which it keeps the Earth and us warm. We measure such by sunspot trends and other proxies or parameters of the Sun's activity, typically radio isotopes extracted from the seabed, ice cores, and tree rings. My previous work on solar activity along with that of Dr. Ole Humlum form much of this subject area.

2. That future hibernations, or global cooling periods, are predictable with a high degree of reliability. A list of some of the many other scientists predicting a new cold epoch are found in this book.

3. That the next solar hibernation has begun, bringing with it catastrophic earthquakes that will soon strike the USA and the rest of the world just as they did during that last solar hibernation over 200 years ago.

4. That everyone in a high-risk earthquake zone within the United States should immediately begin to prepare for these catastrophic geophysical events (CGE). *The worst quakes are expected to strike the United States between next year and the 2030s!*

The structure of this book is based on the following:

1. Presenting initial concepts of the science involved.

2. Examining each of the high-risk earthquake regions of the United States using the same climate template, with carbon-14 isotope data and/or sunspots trends as tools.

3. Knowing when the next solar hibernation takes place, specifically from the remaining years of solar cycle 24 (i.e., now) through solar cycle 25 and 26 (the 2030s and 2040s) and thus constructing a likely timeline when the most damaging quakes may arrive. Each region will be reviewed and its risk by time frame charted out. We will start with the New Madrid Seismic Zone (NMSZ) followed by the U.S. West Coast, including Oregon, Washington, and California, followed by Alaska, South Carolina, Puerto Rico, and finally Hawaii.

4. Establishing that the greatest threat to the United States is the high probability of a multiple catastrophic geophysical events (CGE) or, in this case, highly destructive earthquakes taking place within a decade or two of each other as has happened before during the last solar hibernation.

5. Providing recommendations for individuals, businesses, and the government on preparation for the coming geophysical travail.

Because of time and immediacy of the need to get this book out to those in the United States who will be in the crosshairs of the coming catastrophic geophysical events (CGEs), this book will focus primarily on the United States's earthquake risk even though many of the findings and conclusions discussed herein can be applied globally. It is the Sun after all that is in control, and thus, all citizens of the entire planet will, in their own way, have to deal with the soon-to-arrive record catastrophic earthquakes discussed in this book.

The notification of the government, media, and the general public about the coming cold climate and associated catastrophic earthquakes and volcanic eruptions began during the 2007–2008 period. However, more intensive efforts to alert the people began after 2010 when the SSRC issued a press release warning of the predicted geophysical events. This chronology of notifications is in appendix 2 for those readers who want to dig into the details. Make no mistake; our leaders in Washington, the media, and in the affected state capitals are well aware of the seriousness of the coming catastrophic earthquakes. Sadly, they are all doing little to warn or protect our people despite the strength of the science that has been developed.

Some believe the arrival is already here and began with the March 11, 2011 M9.0 quake and tsunami that struck Japan. Based on the more recent 2014 earthquakes that hit Alaska and Puerto Rico and the level of solar activity when they struck, it does appear the cycle of highly destructive earthquakes has begun in any case. The starting gun has already been fired for the next twenty years of exceptional geophysical stress for our planet and accompanying danger for our people.

Chapter Summaries

To begin *Upheaval!* we will briefly lay out in chapter 1 "Where Are We Today," the latest status for the earthquake threat to the continental United States by the U.S. Geological Survey (USGS). We add in our list of research since 2010 and what we believe the overall threat to be.

In chapter 2, "The Sun-Earth Connection," the reader is provided the essential science of the correlation between the Sun's cycles of activity and their tangible, predictable impact on earthquakes and volcanoes. It is described here with references to key research papers that break down the mystery in the science into easy-to-grasp charts, data, and conclusions about this field of study. This chapter and the entire book, for that matter, have been written for the average man on the street.

With chapter 3, titled "The New Madrid Seismic Zone (NMSZ)," we begin the first of four chapters that deal with an assessment of the several high-risk earthquake zones in the United States. In these chapters, we utilize historical data and information for these high-risk regions as provided by the Federal Emergency Management Agency (FEMA), the U.S. Geological Survey (USGS), regional emergency management offices, and responsible seismic or university organizations. We then develop the IEVPC estimates of the likely time frame for catastrophes to strike each region of the United States. Again, each of these chapters relies heavily on the correlation of solar activity with quake occurrence.

This particular chapter explains the level and schedule for the threat in the New Madrid Seismic Zone (NMSZ) of the central Mississippi valley and the eight states most at risk there. We go into the history of this zone where in the winter of 1811 to 1812 that area saw the most powerful series of earthquakes in recorded history, ranging roughly from M7.3 to M8.0. The NMSZ states are Alabama, Arkansas, Missouri, Illinois, Indiana, Kentucky, Tennessee, and Mississippi. This chapter is placed first in the sequence because the IEVPC analysis of the risk for this region places is at equal or higher threat for catastrophic property damage and loss of life as any area in the USA, even more than some of the better known high-risk quake zones along the U.S. West Coast.

Added effort is devoted to the NMSZ since it now occupies a prominent role in the transfer of goods and materials from one coast to the other and from the south to the north. For example, it occupies the greatest at-risk territory for transfer of oil and gas from the Gulf Coast refineries to the northeastern United States. The predicted loss of these pipelines for an extended period of time would be another collateral catastrophe to add to the direct damage from another NMSZ mega-quake.

In chapter 4, "The West Coast: Widespread Damage," we take a look at the IEVPC's assessment of the latest risk of catastrophic earthquakes based on the solar activity correlation for the two well-known threats, the Cascadia Subduction Zone (CSZ), the San Andreas Fault, and other prominent California faults. In addition, we examine another aspect for earthquake prediction—that of time-based energy flows from deep within the Earth, well researched by Dr. Choi and Dr. Tsunoda. We explain how this additional tool along with solar activity aids us in formulating a threat level and probability for the next period for catastrophic earthquakes.

Chapter 5, "Alaska: Big State, Big Quakes," explores the unique character of the largest U.S. state and why they also have their own special history and future timetable for coming catastrophic earthquakes. The scale of geophysical activity in Alaska sets it apart from the contiguous U.S. states, warranting a separate chapter for just this one state.

Chapter 6 covers the multiple threat areas of "South Carolina, Puerto Rico, and Hawaii." We will address the serious threats facing each of these states and why they should also be given attention in any national plan to prepare for coming CGEs. Because California, our most populous state far and away, receives the most attention from the media and the federal government, these other states are often forgotten. In this chapter, we see why history and the cycles of the Sun suggest they should also be on the watch list. With Hawaii, we found an amazing relationship of earthquakes to solar activity—a complete surprise to us!

In chapter 7, "The Greatest Threat: Multiple Catastrophes," we open up a critical subject. We see effectively no coverage of in the media, and as expected, no discussion whatsoever by our government. In this chapter, we lay out the most likely scenario for the coming catastrophic geophysical period we have now entered. Despite the dire future this chapter predicts, unfortunately, the best available science is leading us to only one outcome: We are likely to enter a new global cooling period or solar hibernation. It will be like that of the Dalton minimum (1793–1830) or worse, the Maunder minimum (1615–1745) the coldest period of the Little Ice Age, lasting from 1350 to 1830. The predicted associated catastrophic earthquake threat starts in 2017 and lasts through the late 2030s.

Based on solar activity cycles as defined by the RC theory, past earthquakes trends, and current deep crustal energy flows, the most likely scenario of natural events for the USA will be one of multiple geophysical catastrophes delivering significant destruction, occurring in quick succession across the nation.

Chapter 8, "Preparing for the Great Upheaval," is the chapter that most people have requested over the years. Because many already accept much of the research surrounding the role of the Sun in climate change and geophysical events, they prefer to quickly get into the subject of how to prepare for a more tumultuous future.

Though this book is about the hard science that supports our warnings, it is nonetheless written for the nonscience members of the community. Out of necessity, some science and new terms, like *solar hibernation or catastrophic geophysical event* (CGE), will be used extensively. Other scientific terms and more detailed research papers are contained in the glossary and related appendices at the back of this book. The essence of the book is placed up front in this introduction and in the first half of the book so that the most relevant information can be read and understood in short order.

To my many new friends who have followed me during the past ten years in my study of climate change, you have my permanent thanks for your kind words, your tangible support, and very important personal encouragement. Hopefully, like *Dark Winter,* you will also find favor with this new book. I believe if you follow its recommendations, you will be better prepared for the coming global upheaval!

John L. Casey
CEO, International Earthquake and Volcano Prediction Center (IEVPC)
President, Veritence Corporation
President, Veritence Publishing, Inc.

Chapter 1

Where Are We Today?

He who is outside the door has already a good part of his journey behind him.
—*Dutch proverb*

We begin the heart of this book with a look at where we are today in terms of broad USA earthquake risk assessments. We will look at the U.S. government's perceptions and then examine what has been said by the authors and other researchers.

Since 2010, we have published research papers on climate and seismology, which in turn we have used to write letters to our government. These letters attempted to convince our leaders in the USA of the need to prepare the country for catastrophic earthquakes expected to strike during the next solar hibernation, the Eddy minimum.

Around 2014, and after many communications with the government, we began to see a positive change in the government's attitude toward the earthquake risk facing the country. It reflected a more serious assessment of the earthquake risk across the USA. Previous studies by the USGS had provided a threat assessment that was tepid at best and likely confusing to the public in terms of whether they should be more or less concerned about earthquakes.

Fortunately, recent studies by the USGS are now more closely aligned with what we have been saying for years. The best yet official vision of the earthquake risk across the continuous USA from their perspective has been published in a 2015 paper, which provided an important update to the earthquake threat. Despite the thoroughness of the recent studies and in some instances realistic risk assessments, they still fall well short of making a case for what the IEVPC believes should be our national action plan to prepare for the predicted era of devastating earthquakes we are facing. This omnibus study was published in June 2015 by USGS researchers K. S. Jaiswal et al.

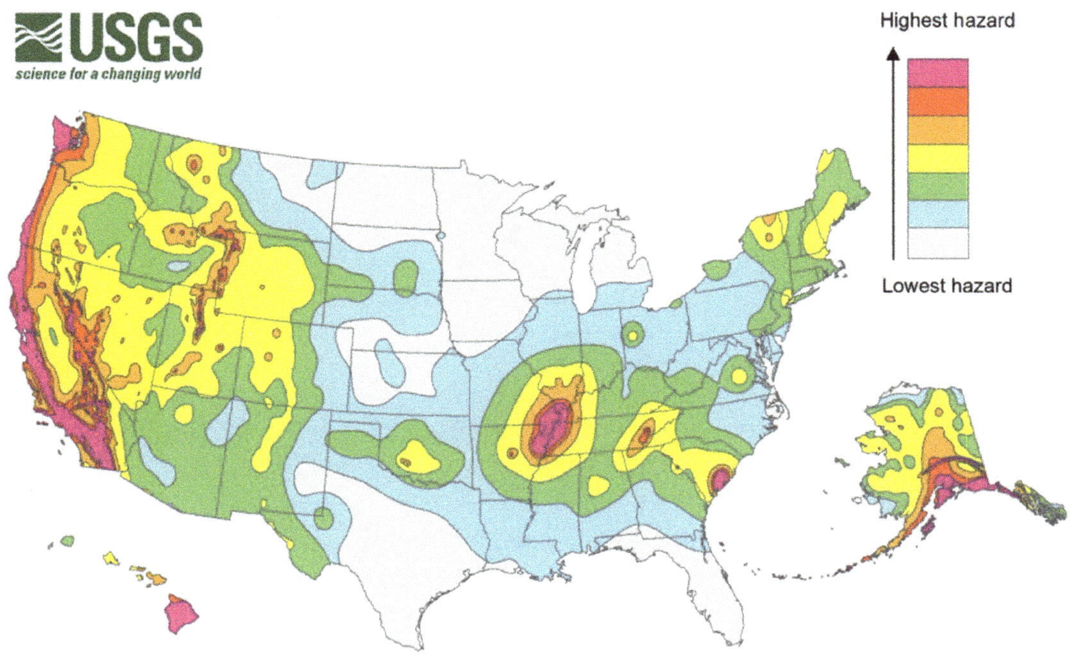

Figure 1. USA earthquake hazard map. USGS map showing the intensity of potential earthquake ground shaking that has a 2% chance of occurring in 50 years. The map uses the Modified Mercalli scale for measurement of the intensity of ground shaking. Source: USGS, Jaiswal, et al.

This USGS study significantly increased the numbers of U.S. citizens who will be impacted by future quakes on the growing U.S. population as seen in the chart below:

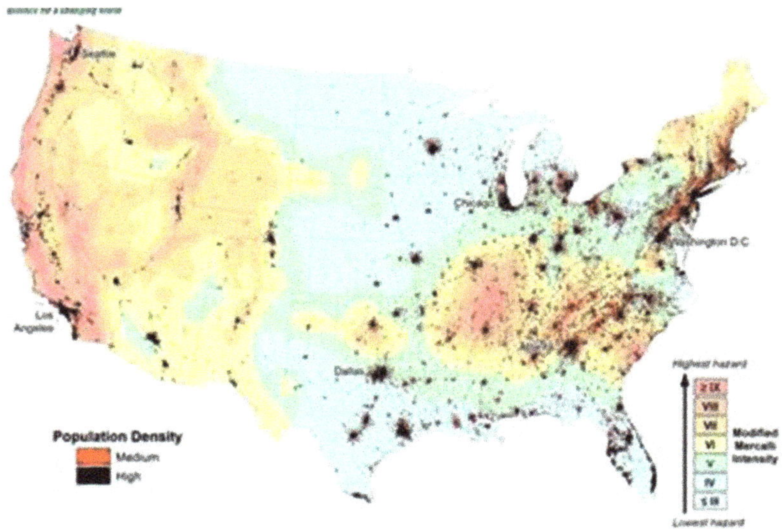

Figure 2. Population density within earthquake hazard areas. USGS map showing (1) the locations of major populations and (2) the intensity of potential earthquake ground shaking that has a 2% chance of occurring in 50 years. Source: USGS, Jaiswal, et al.

Here is the abstract or summary from the study and its important new assessment of the 143 million Americans now at risk from earthquakes. This is double the number at risk from the USGS 2006 assessment.

> A large portion of the population of the United States lives in areas vulnerable to earthquake hazards. This investigation aims to quantify population and infrastructure exposure in places within the conterminous United States that are subjected to varying levels of earthquake ground motions by systematically analyzing the last four cycles of the U.S. Geological Survey's (USGS) National Seismic Hazard Models (published in 1996, 2002, 2008 and 2014). Using the 2013 LandScan data, we estimate the number of people who are exposed to potentially damaging ground motions (peak ground accelerations at or above 0.1 g). At least 28 million (~9% of the total population) may experience 0.1 g level of shaking at relatively frequent intervals [annual rate of 1 in 72 years or 50% probability of exceedance (PE) in 50 years], 57 million (~18% of the total population) may experience this level of shaking at moderately frequent intervals (annual rate of 1 in 475 years or 10% PE in 50 years), and 143 million (~46% of the total population) may experience such shaking at relatively

> infrequent intervals (annual rate of 1 in 2,475 years or 2% PE in 50 years). We also show that there are a significant number of critical infrastructure facilities located in high-earthquake-hazard areas (Modified Mercalli intensity ≥ VII with moderately frequent recurrence interval).

We are pleased with these more realistic earthquake risk assessments. For example, they have increased the likelihood of major quakes in Northern California. This is shown in a March 2015 paper which addressed the threat of a catastrophic earthquake in the San Francisco Bay area. They now believe there exists a:

"72% probability of one or more M ≥ 6.7 earthquakes from 2014 to 2043 in the San Francisco Bay Region." [italics added]

This is almost identical to the estimate we have though with a smaller quake magnitude.

From an August 12, 2015 article in Physics.org, we have the following summary statement about the USGS study by K.S. Jaiswal, et al.:

> More than 143 million Americans living in the 48 contiguous states are exposed to potentially damaging ground shaking from earthquakes. When the people living in the earthquake-prone areas of Alaska, Hawaii and U.S. territories are added, this number rises to nearly half of all Americans.
>
> "The new exposure estimate is nearly double the previous 2006 estimate of 75 million Americans in 39 states, and is attributed to both population growth and advances in science," said William Leith, who is the USGS senior science advisor for earthquake and geologic hazards and a co-author of this study. "Populations have grown significantly in areas prone to earthquakes, and USGS scientists have improved data and methodologies that allow for more accurate estimates of earthquake hazards and ground shaking."

On the other hand, the use of phrases like the earthquake risk is "2% chance of occurring in 50 years" adds nothing to the urgency of the real situation we see from the IEVPC perspective. In fact, it may cause many without a statistics or geology background to conclude the 2% is not much to be concerned about.

For our part, we have been spreading the word about an expected dangerous period of geophysical events for almost seven years. The detailed chronological list of warnings or published science we have made through press releases, books, emails etc., to the media and elected officials and the public are spelled out in a chronology in appendix 2. These communications and past risk assessments made prior to those in this book are extracted here in more simplified reading form:

1. March 2, 2010. The Space and Science Research Center (SSRC) issues its second Research Report 1-2010 (Preliminary) and associated press release. The report is published on the internet for widest possible distribution. The report is titled "Correlation of Solar Activity Minimums and Large Magnitude Geophysical Events." This paper identifies an 80% probability of catastrophic earthquakes and volcanic eruptions during the next solar minimum/hibernation.

2. A paper was published in the as *NCGT Newsletter* entitled "Earthquakes and Solar Activity Cycles" by D. R. Choi and L. Maslov in 2010. In this seminal paper, the connection between solar activity and geophysical events is identified.

3. March 14, 2011. The SSRC issues Press Release 4-2011, warning that there will be more and large earthquakes like that which struck Japan on March 11, 2011.

4. A paper is issued by Dr. Choi and Dr.Tsunoda, validating the previous Choi-Maslov paper from 2010 that linked solar activity with earthquakes and volcanic eruptions.

5. In the fall of 2011, *Cold Sun*, endorsed by other scientists, was published by Trafford Publishing. The book lays out the case for the relationship between solar cycles and record earthquakes and volcanic eruptions.

6. September 13, 2013. The SSRC publishes the commentary/research paper titled "Earthquake/Volcanic Activities and Solar Cycles" by Dr. Choi in its *Global Climate Status Report* (*GCSR*) edition 3-2013. In this paper, Dr. Choi repeats much of the 2010 Choi-Maslov paper, which lays out in substantial detail a strong correlation between the strongest, most damaging earthquakes and volcanoes, and solar activity.

7. March 2014. A summary research paper is authored by Dr. Choi, Dr. Maslov, Dr. Tsunoda, and Casey, which was published in the SSRC *GCSR* edition 1-2014. This summary cites previous research of the authors with the following conclusion:

   The increasing seismic activity since 1990 is expected to continue for the coming two to three decades as we have entered a "solar hibernation" or possibly a mini-ice age (Casey, 2012); this will likely bring more strong, possibly catastrophic earthquakes.

8. April 28, 2014 the president is notified for the *final* time to prepare the nation for the coming solar hibernation via a letter and widely publicized Press Release 2-2014. In the letter, among other items, was the following:

   Research related to these solar hibernations, also shows they occur concurrently with the most destructive earthquakes and volcanic eruptions, the latter of which can add dramatically to an already colder climate."

9. June 2014. Humanix Books publishes an updated, restructured version of *Cold Sun* as *Dark Winter*, which also covered the correlation between record earthquakes and volcanic eruptions with solar hibernations.

10. August 25, 2014. The IEVPC posts and distributes its press release titled "California Enters Greatest Earthquake Risk Period." The release is sent to the Office of the Governor of California and major newspapers in the state. The detailed press release explains the solar activity connection with major earthquakes and the 206-year cycle of the Sun that is also causing a shift into a new cold climate.

11. 2014–2015. Hundreds of radio and NewsMaxTV interviews and public presentations were conducted by Casey, alerting millions of Americans of the need to prepare for the coming cold climate with its associated major earthquakes and volcanic eruptions.

12. Late July, early August 2015. The book *Dark Winter* becomes the number one bestselling book at Amazon.com in the categories of climate change, public policy, astronomy and astrophysics, Earth sciences, and weather. *Dark Winter* warns of a coming cold epoch and associated catastrophic earthquakes and volcanic eruptions.

13. June 5, 2015. The SSRC sends a letter to FEMA Administrator Craig Fugate with a warning that the United States will enter its highest risk period for catastrophic earthquakes and volcanic eruptions during the period of 2017 to 2038.

14. June 2015. Numerous mainstream media outlets and prominent state newspapers in high-risk quake zones were notified via email of the June 5, 2015 FEMA letter and the threat of catastrophic earthquakes.

15. June 2015. The SSRC notified each applicable governors' office of the increased threat of record quakes and volcanic eruptions vis-à-vis the letter to FEMA Administrator Fugate. All West Coast states and central Mississippi states and South Carolina were notified.

16. September 2015. NewsMax Media Inc., parent company of Humanix Books, begins to air a TV documentary about the book *Dark Winter* that discusses the coming cold climate and associated destructive earthquakes and volcanoes. Millions view it on Dish Network, Direct TV, Verizon FiOS, and online at NewsmaxTV.com.

17. May 13, 2016. A follow-up letter and additional warning was sent to the FEMA administrator from Veritence Corporation, indicating that the June 5, 2015 letter from the SSRC should be taken more seriously in view of recent geophysical activity and the warning issued by Dr. Thomas Jordan at the Southern California Earthquake Center on May 4, 2016.

18. May–July, 2016. A second mailing is sent out to the governors of the U.S. States most likely to suffer the brunt of coming geophysical disasters.

19. June 6, 2016. The IEVPC posts and distributes its first press release of the year, "Federal and State Leaders Warned to Prepare for Catastrophic Earthquakes and Volcanoes." This release, as indicated in the text, represents the final warning to governments, after years of prior warnings, to get their states and federal agencies into immediate preparation mode for catastrophic quakes that could strike anytime but definitely within the period 2017 to 2038.

As we will see in the remaining chapters of this book, although there have been many warnings, our country remains substantially unaware and therefore unprepared for the dangerous years ahead.

Chapter 2

The Sun-Earth Connection

Every new truth begins in a shocking heresy.

—Margaret Deland

It did not take long after the discovery of the 206-year solar cycle and the formulation of the relational cycle theory of climate change, or the RC theory, before it was also clear that the Sun had much more impact on the Earth than just the diurnal cycle, climate, or the change in the seasons.

Figure 3. The Sun. Source: NOAA, January 19, 2005.

In this chapter we examine both the means by which we measure the Sun's activity and look at several research papers that have exposed the relationship between the Sun and geophysical process on Earth in the form of earthquakes and volcanic activity. The detailed papers supporting this explanation are found in their original form in appendix 2.

Prior to jumping into the science behind this Sun-Earth or solar-terrestrial connection, it is important to understand how scientists measure the Sun's behavior and climate change over long periods of time. As has been mentioned earlier in this book, a primary element to the assertions made herein is that climate changes, especially to cold climates, are a key component to determining when our worst earthquakes and volcanoes happen. Therefore, we need methods for analyzing when past climate changes took place to see if a correlation exists. These climate changes, going back many hundreds of years provide for us a template for understanding the Sun's past behavior, giving us the ability for estimating future activity for both the Sun and geophysical events. Measurement of the global temperature changes then becomes the key to the process. How do we do that, and how far back in time can we go in measuring the Earth's temperature changes?

We start at the year 1979. That is when the U.S. government first began to monitor the Earth's temperatures via satellites. We have detailed global temperature records from that time to present. But that is such a small period of time. Temperatures prior to that were measured directly from thermometers but only back to 1850 here in the United States. European temperatures measurement went back beyond that to the 1600s. This is also the time when Galileo Galilei began to use the telescope to observe sunspots on the Sun. Therefore, since 1610, roughly, we have had a second means of measuring the Sun's activity level—sunspots.

Going back further in time before mankind began to measure temperature or the Sun's energy level, we are required to find proxies or analogous indicators to temperatures. The science surrounding how this is done is extremely interesting and is worthy of another book, but for this one, we will be brief.

The most common means of determining the Suns' activity levels over the millennia has been in the use of isotopes of atoms which are created during various cold or warm phases of the Earth's climate. The most used among these isotopes, carbon-14 (C14), oxygen-18 (O18), and beryllium-10 (Be10). They are extracted from various sources like deep sea floor (benthic) core samples. They are also taken from ice cores in the Antarctic, most notably from the Russian Vostok station, or from ice cores taken from Greenland.

In this text, and in addition to the use of sunspot measurements, we have relied on the C14 records. Vital to this isotope's use has been the accumulation of a highly reliable data set made of a variety of C14 sources globally. The importance of this data set was discussed in *Dark Winter*. One of the finest pieces of basic research in C14 has been our choice, the INTCAL 2004 C14 data set compiled by Dr. Paula Reimar, et al. With the proxy information in hand, many scientists worldwide have been able to correlate a host of natural and geophysical events to past global temperatures and the Sun's behavior.

That C14 data was used to compose the following basic chart of solar activity going back to the year AD 445. The entire INTCAL 2004 data goes back 24,000 years. In order to not make this book too cumbersome and to focus on recent trends of geophysical activity, we have limited our perspective to the last 1,500 years approximately. It is from this chart and the data behind it that the cycles outlined in the RC theory were first discovered.

Here is the C14 chart, unadorned, from AD 445 to 1950.

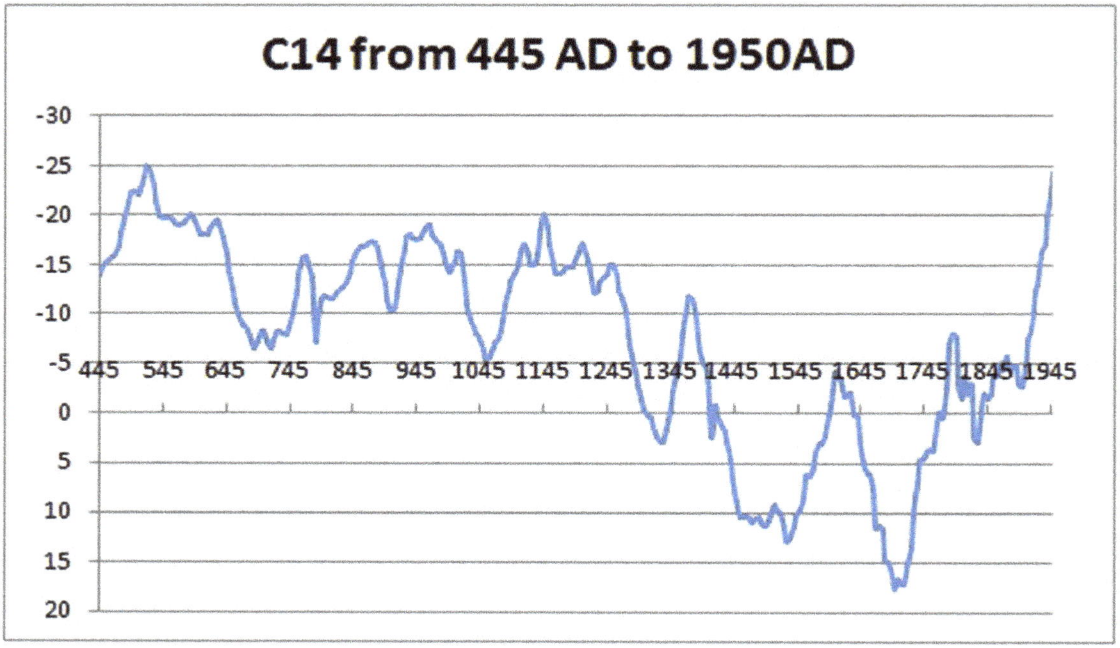

Figure 4. Carbon-14 measurement from AD 445 to 1950—a proxy for solar activity. Source: J. Casey, Data, INTCAL 2004, Reimar, et al. The solid blue line shows the relative warm climate periods (negative numbers of the x-axis) or cold climate periods (positive numbers). Note the chart ends in 1950. The blue line takes off after that because of both the Sun-caused warming of the modern warm period and the use of atmospheric atomic bomb testing done by the USA and the former Soviet Union. The latter caused an artificial though marked jump in the C14 count in the atmosphere.

Where the C14 falls short in recent years after 1950, we have detailed sunspot measurements and actual temperature measurements to complete the solar activity history from which we can then compare earthquake activity to decide whether there is in fact a meaningful correlation between them. So our next step was finding a good set of sunspot data to complete our climate history picture. Again we do that using historical sunspot records back to the early 1600s. Below is the sunspot chart produced by Dr. Ole Humlum, a coauthor of this book and a professor of physical geography at the University of Oslo, Norway. Dr. Humlum, also is a glaciologist and geomorphologist who was also the coeditor for the previous Global Climate Status Report (GCSR) from the SSRC. In the GCSR, 24 different parameters of solar system and planetary activity were routinely monitored to determine which way the Sun was going on the matter of climate change, i.e., global warming or global cooling. The SSRC established itself as one of the most accurate climate prediction companies in the USA. By the way, CO_2 was never used as one of the parameters followed in the Global Climate Status Report since the SSRC determined CO_2 had no major role in climate variation!

Throughout this book, we are indebted to the thorough, detailed, and objective treatment of climate science by Dr. Ole Humlum. Without his many charts and behind-the-chart data compilation found in this work, our book could not have been written. Here is one of his pieces—the sunspot chart and template used in this text in establishing a geophysical correlation to the Sun's behavior:

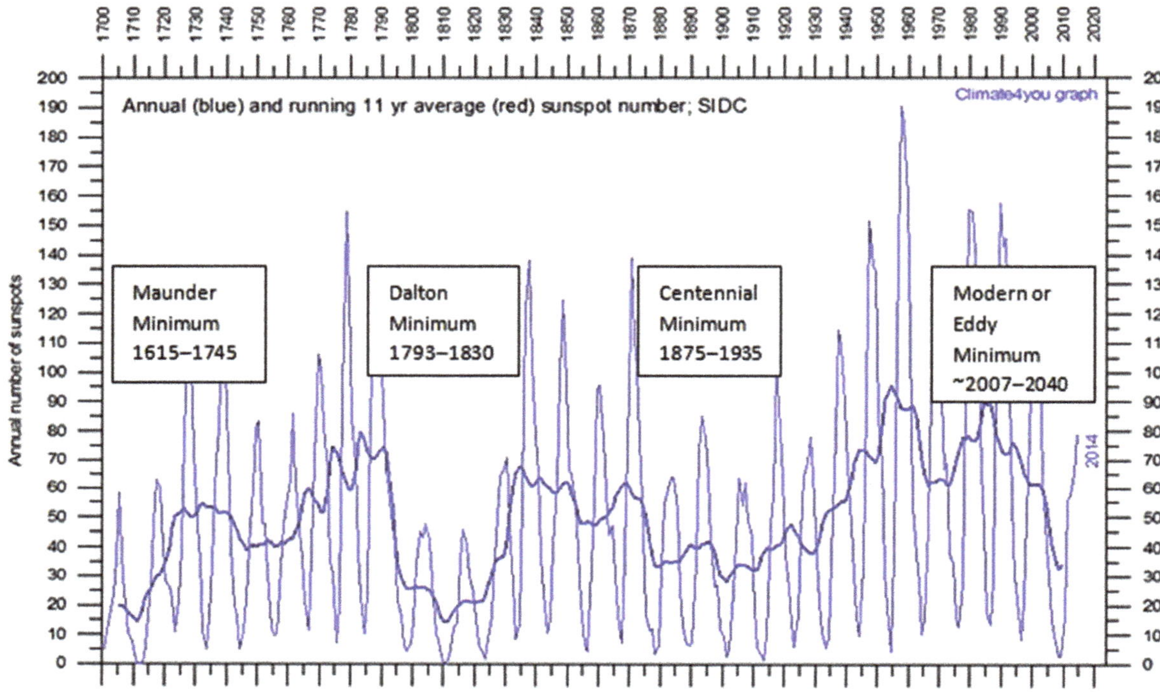

Figure 5. Sunspot counts from 1700 to 2014. Source: Dr. Ole Humlum, Climate4you.com. The heavy blue line denotes the smoothed average of each 11-year sunspot cycle. The thin blue lines demark each 11-year solar cycle. The four labeled periods of major declines in solar energy output (the Sun) are indicated. The step drop in solar activity centered around 1970 is called the multicycle pause.

The term *solar minimum* is synonymous with *solar hibernation* for those periods when the sunspot count drops below 50 at the peak of any 11-year solar cycle. The centennial minimum is a significant decline in solar energy associated with the centennial cycle noted in the RC theory. The low points or the bottoms of these minimums are identical to the coldest periods in terms of global temperatures. The Eddy minimum is the recent designation by the solar physics community in honor of the late Dr. John Eddy who did pioneering work in solar activity and sunspot research.

As one can see, the Eddy minimum probably started its decline when the average sunspot count (dark line in the chart) reached the same low level as the centennial minimum centered on the year 1900. In the last few years, we have seen 100-year cold temperature records broken around the world during winter. This decline in solar activity and solar energy by which the Sun warms our planet is a primary reason why we have witnessed record cold periods.

Another means for evaluating the Sun's energy output is through the use of measurements of total solar irradiance or TSI. The chart below reflects the amount of energy delivered at the top of the Earth's atmosphere coming from the Sun. This amounts to 1,366 watts per meter squared as measured by a set of satellites. Newer satellites use a standard of about 1,361 watts per meter squared (W/ms^2).

This chart is provided here for understanding several important issues though we will primarily reply on C14 and sunspot charts in this book:

1. TSI is a highly accurate measurement of global temperature trends.

2. Only a few watts (~3) reduction in the Sun's output can mean the difference between a Maunder minimum like during the Little Ice Age and the modern warm period.

3. The current trend of TSI for the Earth shows a dramatic drop in TSI has begun certainly since 2007 (far right of the chart). This is but one of many indicators that reinforces a prediction for a coming cold climate and a potentially dangerous one in its global impacts for food production and social stability.

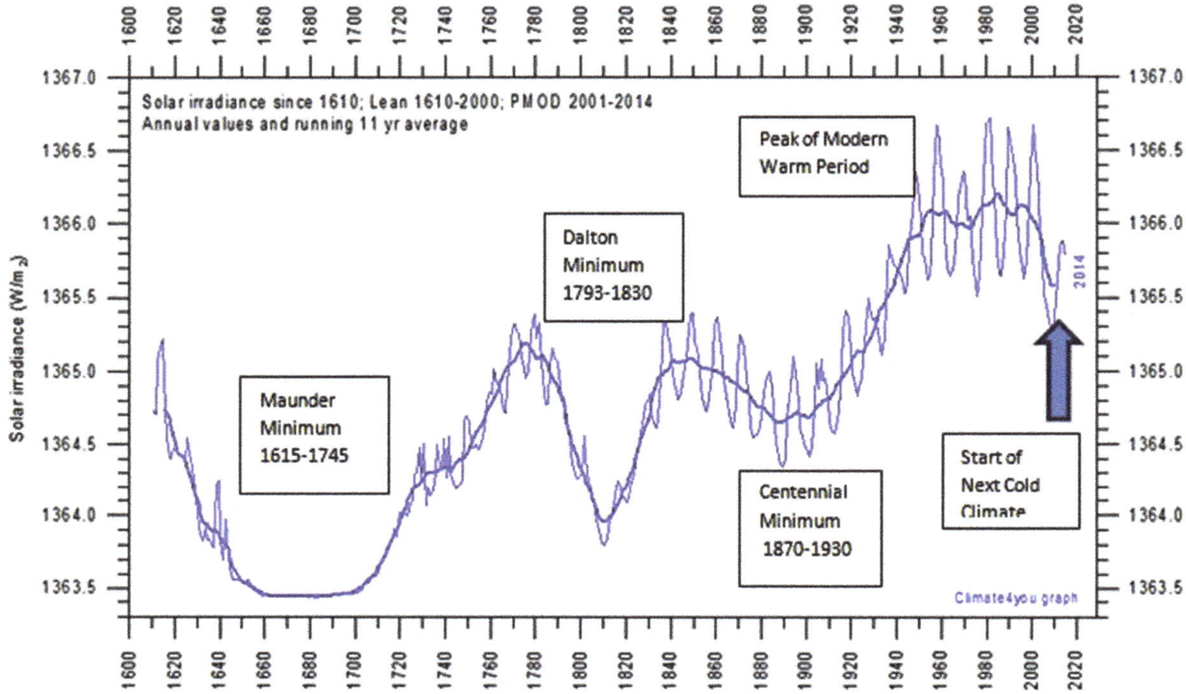

Figure 6. Total solar irradiance (TSI) measurement from 1600 to 2014. Source: Dr. Ole Humlum, Climate4you.com, labeling, J. Casey. Note the significant drop in the energy by which the Sun warms the Earth—shown at the far right of the chart. The next cold epoch has begun. Sunspot minimums are shown for comparison. Sunspot minimums are in sync with TSI output from the Sun as one might expect. Note the very small drop of only 2.5 watts between the peak of the modern warm period and the bottom of the Little Ice Age which was seen during the Maunder minimum.

What is the Sun doing in our current 11-year solar cycle number 24?

Here is the latest chart of sunspot activity from the Sunspot Index and Long-term Solar Observations (SILSO) in Belgium, a leading solar research center and one of the oldest tracking facilities for the Sun's behavior.

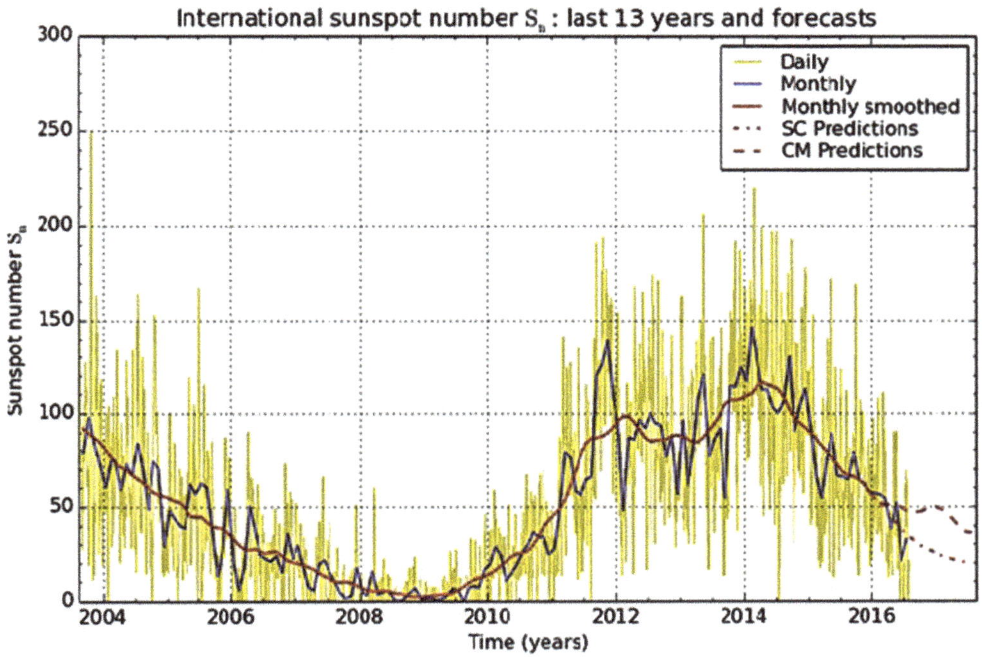

Figure 7. Solar Cycle number 24 sunspot trend. Source: SILSO, Belgium. Data through August 1, 2016.The average sunspot curve is shown in red with the daily sunspot numbers shown on the varying blue line. SC and CM are two different techniques use by SILSO for prediction of future sunspot activity.

Several items of interest are present in this SILSO sunspot chart:

First, we see the last half of solar cycle number 23 to the left of the chart and the slow decline in sunspots that led to the normal low point in sunspot counts between solar cycles, centered around 2009. This period of near-zero sunspots was exceptionally long for a transition period between cycles and was an early indicator that cycle 24 would be a weak one in terms of sunspots and other solar activity parameters.

Next, we see in cycle 24 that it had the two typical peaks in sunspot activity near the 11-year-cycle high point of activity. The special situation here is quite important. The second of the two peaks is the most powerful, reaching an apex in 2014. This is a unique signal of a coming cold epoch that was announced in one of the last climate reports published at the SSRC. This only happens in a long-term downward trend of multiple 11-year cycles of the Sun and then only before a cold climate begins. This spike in activity was followed by a corresponding jump in global temperatures in 2015 and 2016 at the same time the normal 4-year El Niño cycle was nearing peak. This combined effect led to a brief though near record global temperature jump—a situation only explainable by solar activity and one completely unexplainable by man-made global warming rationale. Again, the greenhouse gas theory does not work.

Finally, one can easily see that a steep drop in solar activity (sunspots) has begun after the peak of cycle 24 as we should expect when heading into a new cold epoch postulated by those listed in the back of this book. Anticipating this event led to the posting of an April 2016 climate predictions in the "Commentary" pages at the web site of Veritence Corporation (www.veritence.net). The commentary was titled "Peak Global Warming" and had a prediction for 2016 temperatures and a prediction for long-term planetary temperatures hundreds of years ahead. The first prediction appears to be spot on as the May and June 2016 global temperature drop was the steepest in almost 4 decades according to satellite temperature measurements from Dr. Roy Spencer.[4] More declines are to follow.

It is now widely assumed within the solar physics community that the next two 11-year solar cycles, cycle number 25 and cycle number 26 are expected to have, at peak, an average of 50 sunspots. This number compares with the last few solar cycles up to the declining cycle 24 that saw sunspot numbers over 150 at peak. These next cycles, by definition, at that low 50 sunspot level of solar activity, would then constitute a solar hibernation or grand minimum. Our calculations are such that a hibernation of the type of the Dalton minimum of 1793 to 1830 with two cycles of 50 sunspots each are likely as NASA now believes. It should be noted that other researchers believe an even colder climate is approaching and would reach the same depths as the Maunder minimum at the bottom of the Little Ice Age (LIA). The broader Little Ice Age that helped cause the demise of the Vikings' Greenland settlements, founded during much warmer times, extended from 1350 to 1830 in general, encompassing the Wolf, Sporer, Maunder, and the Dalton minimums.

The bottom line from this discussion of solar activity is this:

A new multi-decade long cold climate is rapidly approaching consistent with solar activity cycles as defined by the RC theory. The coldest first year of the bottom of the cold period is 2031 though significant cold weather effects will occur before that year and throughout the 2030s if not the 2040s.Though the next cold climate is predicted to be as cold as the Dalton minimum from 1793 to 1830, there is a credible risk that the new cold epoch could be as dangerously cold as that seen during the previous Maunder minimum (1615–1745) at the bottom of the Little Ice Age.

****

Now that we have the big picture on how we measure the Sun's behavior with climate changes, we can look into whether there is in fact a strong connection between the Sun and geophysical processes on Earth, like major earthquakes.

One of the first papers trying to understand the correlation between these periods of low activity on the Sun as measured by sunspots was published as the SSRC Research Report 1-2010, March 2010. The full paper is in appendix 4. Seven years later and the paper is still just as valid as when first made public and posted online for the whole world to critique. Here are the copied abstract and conclusions from that paper.

> Abstract
>
> An independent review of historical records was performed for 350 years of global volcanic activity (1650–2009) and seismic (earthquake) activity for the past 300 years (1700 to 2009) within the continental United States and then compared to the Sun's record of sunspots as a measure of solar activity. All three data sets were examined to determine whether a relationship existed between them and if the results of such a study could be used to develop methodology for identifying future geophysical events. The preliminary results from the study have shown that there exists a strong correlation between the solar activity that causes climate changes and the Earth's largest seismic and volcanic events. The impressive degree of correlation for global volcanic activity (>80.6%) and for the largest USA earthquakes (100% of the top 7 most powerful) vs. solar activity lows provides a basis for future estimates of the time periods and magnitudes for the largest volcanic and seismic events many decades in advance. Finally, the coincidence of the Centennial and Bi-Centennial cycles of the RC Theory showed unmistakable relationships to these largest geophysical events. The use of such a tool may provide a new and valuable method for protection of people and property located in and around high risk geologic zones. Further, a

significantly increased risk is indicated during the next 20 years for volcanic and earthquake events of historic scale.

Conclusions

[6] As a result of research conducted, it is reasonable to conclude there exists a strong correlation between global volcanic activity among the largest of classes of eruptions and solar activity lows. With the 80.6% occurrence of large scale global volcanic eruptions taking place (>VEI 5) during solar activity lows and with 87.5% occurring for the very largest (>VEI 6) eruptions during major solar minimums, it is concluded that any reliable predictive tool for forecasting future solar activity would also lend itself to forecasts for future global volcanic eruptions of the most powerful magnitudes. For example the RC Theory of solar activity may be an effective tool for forecast of global volcanism.

[7] The occurrence of each of the largest seven USA earthquakes during solar activity lows and in particular during solar hibernations indicates a predictive tool like the RC Theory for future extended solar minimums may also be effective in forecasts of major USA earthquakes.

[8] Given the unusually high degree of correlation found in the study for both the highest levels of global volcanism and USA earthquake activity when compared to extended solar activity lows, it can be concluded that there exists a significant likelihood (greater than 80%) that the current recently started solar hibernation may result in historic scale global volcanic eruptions and record earthquake activity within the continental United States.

****

Because of the results of this paper and those that have followed over the intervening seven years, the authors believe the United States should effectively be on a war-time footing to get our people ready for the catastrophes that are just around the corner. Again, to restate this important opinion:

"It can be concluded that there exists a significant likelihood (greater than 80%) that the current recently started solar hibernation may result in historic scale global volcanic eruptions and record earthquake activity within the continental United States."

****

> [9] The determination that solar activity cycles may indicate timing and intensity of geophysical events like volcanism and earthquakes points toward a possible connection between solar activity and the underlying cause of these geophysical events, namely plate tectonics.
>
> [10] The solar hibernation identified by Casey (2008) is currently under way. The results of this study and the high correlation between described volcanism and earthquakes and solar hibernations warrants the widest dissemination of warnings to personnel and governing organizations in high risk geophysical zones. It is expected beginning at any time and during the next twenty years of the solar hibernation, that potentially historic volcanic eruptions are likely globally and similarly record setting new earthquakes are likely within the continental United States.

The 2010 paper may have been the first that quantified the risk of major earthquakes and volcanoes during the cold phase, the solar hibernations, of the Sun's cycles.

In subsequent letters to FEMA in June 2015 and May 2016, research was explained that made it clear that the probability for a significant geophysical risk existed and the window of opportunity for these risks to reach critical stage was 2017 to 2038. This is still a reasonable period for a likely catastrophic strike for those areas of the United States with known major earthquake zones.

It is a stunning revelation that the Sun's cycles has such a strong correlation to geophysical processes on Earth. This table below also extracted from the paper explains how the seven largest earthquakes in U.S. history came during a solar activity low point, i.e., solar minimums or solar hibernations:

Top Seven Largest Continental USA Earthquakes (Source: USGS*)

Location	Date	Magnitude	Associated Solar Minimum
1. Cascadia Subduction Zone	01-26-1700	~9	Centennial: Maunder
2. New Madrid, Missouri	12-16-1811	8.1	Bicentennial: Dalton
3. New Madrid, Missouri	02-7-1812	~8*	Bicentennial: Dalton
4. Fort Tejon, California	01-09-1857	7.9	Intermediate Minimum***
5. San Francisco, California	04-18-1906	7.8	Centennial Minimum
6. Imperial Valley, California	02-24-1892	7.8	Centennial Minimum
7. New Madrid, Missouri	01-23-1812	7.8	Bicentennial: Dalton

*Measurement methods vary. The USGS says the New Madrid Feb 7, 1812 temblor may have been 8.8 on the Richter scale.

**Centennial and bicentennial cycles from the RC theory have periods of 90–100 years and 206 years respectively.

***Intermediate minimums are easily observed declines in solar activity (sunspots) though lesser in magnitude than centennial or bicentennial events.

****

The magnitudes of the New Madrid quakes have been scrutinized numerous times by various researchers and the USGS. The generally accepted magnitudes today versus what was known when the 2010 paper extract above was written have since changed according to the USGS. They now believe there were three main quakes and a major aftershock—all in the M7.3 to M7.5 range. More about these massive series of quakes are examined in the chapter on the New Madrid Seismic Zone with the estimates of other researchers who believe multiple M8.0-plus quakes may have taken place.

As discussed in *Dark Winter*, the centennial and bicentennial cycles noted in the RC theory had been previously found by others and named the Gleissberg (90–100-year cycle) and the Suess or DeVries cycles (206-year cycle), respectively. For a variety of reasons, these previous scientists' discoveries never became public information except within the professional journals followed by a narrow community of scientists who were interested in such research.

The next paper that highlights the Sun-Earth connection was a joint paper by Dr. Maslov and Dr. Choi. Both of these scientists had their peer-reviewed paper published in the *New Concepts in Global Tectonics Newsletter*, no. 57, December 2010. The newsletter later became a journal. Even today, this thoroughly researched paper and its breadth of findings appear a special one in many ways. The full paper is also in appendix 4. Here are some of the most import elements and findings from this thirteen-page paper:

a. The researchers examined earthquakes in the range of magnitude M5.0 to M5.9 and found what I have called the Choi-Maslov relationship. In effect, what was discovered was that there is an inverse relationship between moderate size quakes and the Sun's activity level as measured by sunspots. It is depicted by the chart below:

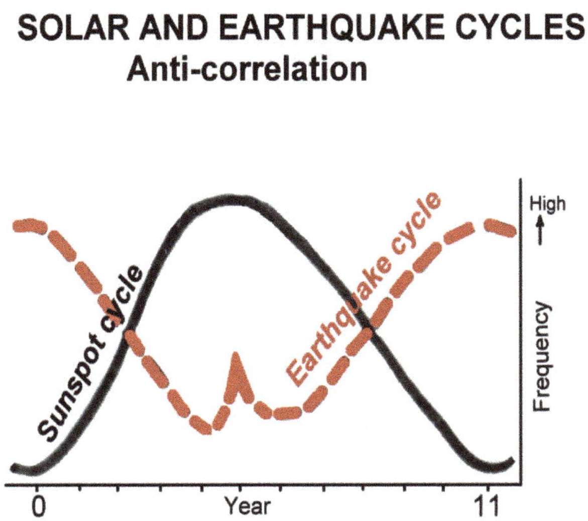

Figure 8: The Choi-Maslov relationship, the existence of an anticorrelation between sunspots and earthquakes.

In Figure 8, we see the Choi-Maslov relationship showing the inverse situation that exists between sunspots and moderate quakes (M5.0–M5.9). The second important finding of the Choi relationship is that at the peak of sunspot activity for the average eleven-year solar cycle, there is a spike in earthquake activity. In 2014 we had our first major earthquakes of the new cycle of earthquakes, one in Puerto Rico (M6.4) and another in Alaska (M7.9) right at peak of solar cycle 24. These findings are two of the most important elements of a potentially new field of science that clearly demonstrates the strong linkage that exists between the Sun at 93 million miles away and the very movement of the Earth's —truly a foundational discovery.

Dr. Choi and Dr. Maslov's 2010 paper was published again in the SSRC's Global Climate Status Report (GCSR) of September 13, of 2013. Dr. Choi's commentary, titled "Earthquake/Volcanic Activities and Solar Cycles," laid much of the foundation for later papers.

b. The full commentary is at appendix 2. The paper also described a quiescent period from 1996 to 2003 where global earthquake activity declined substantially—a unique occurrence which warrants further research. Here, though, is the important final conclusion paragraph from Dr. Choi's commentary and the Choi-Maslov 2010 paper:

> **4. Conclusion**
>
> This paper introduced how the Sun and other powerful planetary forces are intricately interacting with the Earth's geodynamic events, represented by earthquakes and volcanic eruptions. Long-term solar cycle trend is most crucial in understanding the past global climate, and hence in predicting future climate trend. As advocated by Casey (2012 and 2013a & b), it is doubtless that the Earth has entered a major low cycle or "hibernation" era, and we will have more catastrophic earthquakes and volcanic eruptions in the coming several decades.

By the end of the third year of the next solar cycle 25, around 2022, we will have enough data to accurately say whether we have entered a Dalton-class minimum or a Maunder-class minimum. Given the biblical scale of destruction a Maunder Minimum cold climate, observed at the bottom of the Little Ice Age, would produce in terms of the global loss of food from cold damage to crops and resultant loss of life, we should all hope that a Dalton-class solar hibernation is coming! That will be difficult enough.

These various assessments of the sensitivity of the Earth's climate to the Sun's variability produces the following opinion:

"Not only do we live in the Goldilocks zone in our revolution around the Sun where we are not too hot and not too cold, but that we live on a knife's edge within that Goldilocks zone. The slightest variation of a fraction of 1% in the Sun's effective TSI on our planet can, make the difference between a modern warm period as we have been living through, and a new 100,000 year global ice age."

Those who have read *Dark Winter* (2014) are familiar with the next charts showing the mathematical curve of the next solar hibernation where the 206-year cycle from the RC theory is the primary driver of the current climate change to a new cold epoch. The bottom year of the cold phase is easily pulled off these charts as the year 2031. Using that year as a reliable benchmark and under the belief that a Dalton-class hibernation is en route, we can draw out several forecasts for the next 20–30 years of the Earth's climate and associated geophysical threats. In turn, using the correlation with historic records for earthquakes and volcanic eruptions, we can estimate the highest risk period of future catastrophic earthquakes.

Based on this analysis, the following two charts were drawn up showing the next bottom of the cold phase of the recently started hibernation will be the year 2031. Using that date, we can then estimate from past quakes and low sunspot years of those past hibernations when our period of highest risk will be. That period was found to be 2017 to 2038.

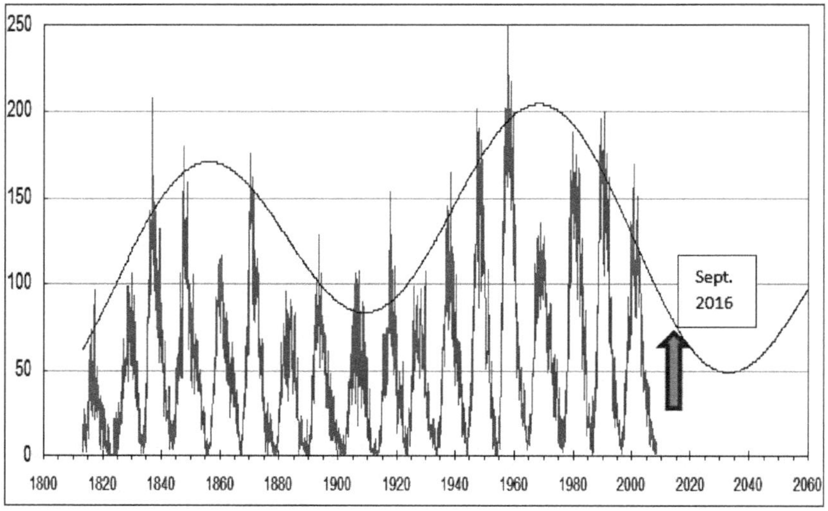

Figure 9. The 206-year cycle and the next solar hibernation. Source: John Casey/M. Vuckcevic.

In this chart above, we see the past 206-year cycle bottoms in the years centered on 1800, followed by another small dip with the centennial or Gleissberg cycle around 1900 and the next bottom calculated for 2031. Though this chart was with data through 2010, the current solar cycle 24 (not shown) has stayed within the calculated sunspot curve.

A closer view of the next solar hibernation displaying the period from 1960 to 2060 is shown in the following chart.

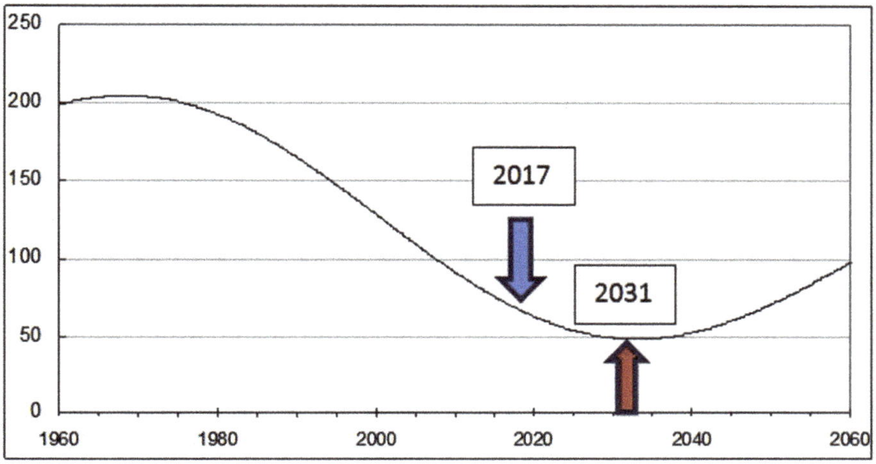

Figure 10. The predicted next solar hibernation. The blue arrow shows where we are in time on the 206-year cycle curve. The red arrow indicates where the calculated bottom of the next solar hibernation is in the year 2031. Note, solar activity stays low for all of the 2030s under a Dalton-class minimum. Under the relational cycle theory of climate variation, the mathematical curve above displays the general timeline for the predicted cold phase or solar hibernation of the Sun. We do not start to warm up from this cold phase until the 2040s and then only slowly. Source: J. Casey, M. Vucevic.

Given this scenario and the historically bad outcomes that it brings, you may be thinking what if the GHG theory is correct after all, and there are no forthcoming cold climates, i.e., no earthquake threat?

Here's the answer to that question. The U.S. government and the UN, after spending billions of dollars, have developed over 70 atmospheric and oceanic global climate models (AOGCM) over the past decade or two. These models based primarily on mankind's production of CO_2 form the central rationale for their predictions of global warming by the year 2100, of increasing sea level rise, and a host of other man-made global warming scary scenarios.

Fortunately, according to an analysis done by Dr. John Christy and Dr. Roy Spencer at the University of Alabama–Huntsville, we know these models have been in error by a huge margins and from a scientific standpoint are completely unreliable. See the chart below. From this chart, we see that the greenhouse gas theory, which has been a weak theory since it was first made popular in 1896, not only remains weak but also the expensive climate models of the manmade global warming community are spectacular failures in predicting climate variation. Most have error rates near or over 200% when compared to the average of all the models versus the actual global temperatures.

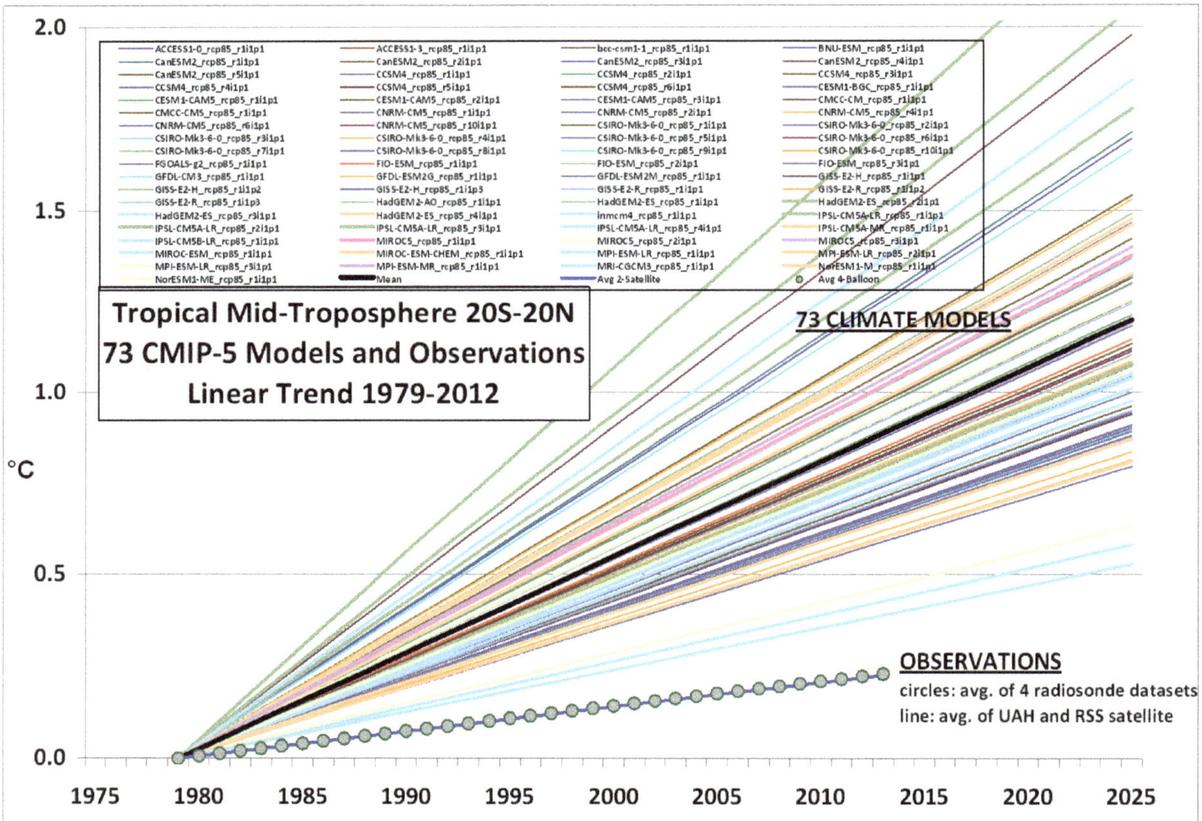

Figure 11. Comparison of 73 global climate models under the greenhouse gas (GHG) theory vs. actual temperatures. Source: Dr. John Christy/Dr. Roy Spencer. In this chart, we see that the GHG climate models when averaged (black solid line) show error rates of 200% with one model at almost 600% over actual measured temperatures (lower dotted line) from radiosondes and satellites between 1989 and 2012.

The final actual measured temperature in this chart is about 0.25° C (in 2012) in global temperature growth since 1980 shown in the dotted line. The first important observation from this outstanding piece of climate model analysis by Doctors Christy and Spencer is that none of the UN or U.S. government 73 models came close to predicting actual global temperatures. The average of the 73 global climate models (solid black line) based on the greenhouse gas theory shows a huge difference of about 0.75° C. That's a 200% error from actual temperatures (0.75 – 0.25/0.25 x 100% = 200%). The most heated among the greenhouse-gas-based climate models shows an estimated temperature of about 1.75°C between 1980 and 2012. That corresponds to an error of temp estimate of 600% (1.75 – 0.25/0.25 x 100% = 600%)! The three "best" models are roughly 80% off actual global temperatures!

Keep in mind that scientists get downright fearful when they have errors 10%, 20%, or 30% off their predicted values for validating a theory. But errors over 80% or over 200% and as much as 600%! As scientific tools, these many models based on the greenhouse gas theory are clearly worthless. And therefore, so is the greenhouse gas theory as a tool for predicting global temperatures, sea levels, glacial ice melting, the numbers of polar bears, or any aspect of climate change tied to the greenhouse gas theory and CO_2 in the atmosphere.

In summary, our situation worldwide and certainly in the USA, is this:

1. A new cold epoch, a solar hibernation, has already started.

2. Past solar hibernations are strongly associated with our worst earthquakes.

3. Since the Sun is the primary driver of climate change, we should expect a repeat of the climate conditions and collateral geophysical and human impacts we saw during the last solar hibernation.

Chapter 3

The New Madrid Seismic Zone (NMSZ)

Those who cannot remember the past are condemned to repeat it.
—George Santayana

The quote above from George Santayana is the same one included in *Dark Winter* and is once more included in this book—this time out of burning necessity!

It is even more relevant now. A repeat of history, both natural and human, is about to deal with us all in way that no one alive today has experienced—and it isn't good news. Nature appears ready to deliver a species-threatening combination of disastrous cold climate in conjunction with record earthquakes and volcanic eruptions that will be the worst in over 200 years.

Arguably, because of the corruption of climate science by the U.S. government and the UN-IPCC, we will be unprepared for both the new cold epoch and associated record earthquakes. Tragically, we will have failed once again to remember the past.

The vast majority of us normally pay no attention to history as we scurry about trying to get a job, trying hold a job, make a family, and trying to hold on to both. We are dominated on a daily basis with mundane matters such as paying bills, medical issues for the young and old of us, and decisions like whether to pay the auto repair shop another $800 to keep the junk in the driveway running a little longer or whether to pick up a new car and associated debt. Much of life's routine challenges are laid out for us, consuming our day with avarice. Who has time for a history course?

In fact, many of us have developed a lifelong adverse reaction to the word *history*. This unfortunate outcome is probably the result of our primary grade education system that forced us to memorize endless lists of dates and events which in turn were regurgitated each Friday to score well in weekly history class tests. Such a deleterious academic practice has contributed to the present ignorant state of history learning among high school and even college graduates. As referenced earlier in this book, how many times have we seen man-on-the-street interviews on TV where those questioned don't know who our founding fathers were, much less crucial decisions they made that created the Constitution and United States of America. This sad situation is even more lamentable when the general populace is queried on what was learned from past events and whether similar situations today can be compared to these past times in U.S. and global history. The proof is everywhere—most of us simply ignore the lessons of the past.

Yes, human history repeats. So, we are about to see, does the natural world.

There is a central theme in this book that links both the natural history with humanity's history. If we study past events, we have the opportunity to change things this time, to get prepared based on lessons learned.

We are essentially history illiterate as a populace. Thus, we are left dependent upon our scientists, media, and especially the government to safeguard us and, as a minimum, inform us of what natural threats we are about to encounter. What happens, though, when these sources say nothing about what's coming our way with potentially disastrous consequences? Or worse, what happens when they cover up what's coming as our current slate of leaders in Washington, D.C., and the mainstream media appear to be doing?

The following history from the last solar hibernation of the Sun in the central Mississippi valley will help give a hint of what to expect—what our current US government is not sufficiently preparing us for. Here is the chart that shows why everyone in the central Mississippi valley should be preparing immediately for the coming devastation. Everyone living in the eight states that surround the NMSZ should see this chart and be told why it's vital to their safety. As they say, this picture says a thousand words:

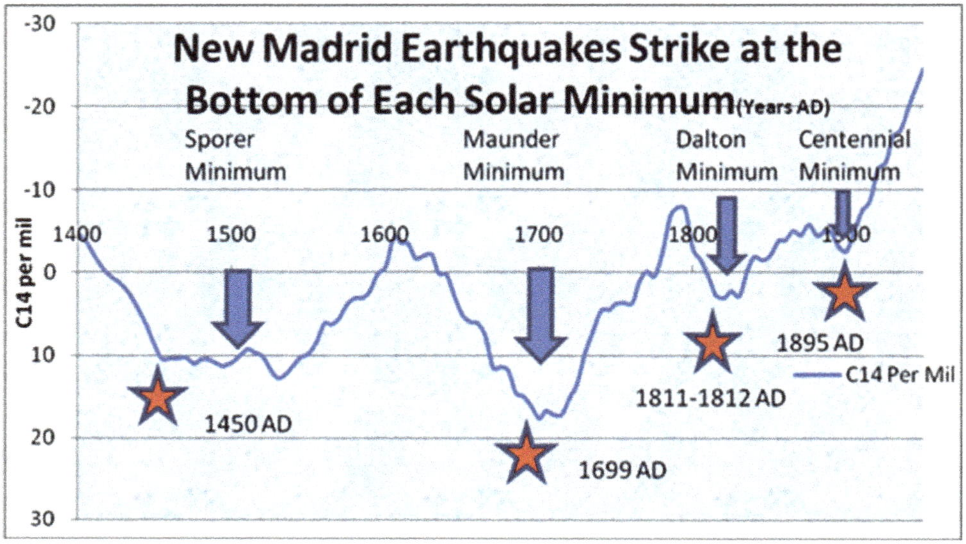

Figure 12. The correlation of past catastrophic earthquakes with solar minimums for the NMSZ. This chart tracks global temperatures using carbon-14 as a proxy for temperatures. Parts of four solar minimums, including three solar hibernations, are shown with deep cold climates between the peaks of the cycles. Quakes (red stars) with estimated magnitudes are as follows: 1450, M7.0-M8.0; 1699, M7.0-M8.0; 1811 to 1812, M7.3, M7.5 plus one at M7.5-M8.0; 1895, M6.8. Source: Casey, data; INTCAL2004 and the USGS.

Note in the figure above, the cold phases are shown as the Sporer minimum, the Maunder minimum, the Dalton minimum, and the centennial minimum. Note also the peak of warming at roughly 1400, 1600, 1800 and the current peak around 2007, demarking elements of four 206-year cycles. As discussed in the previous chapter, the cold phase or solar hibernation has begun now that solar cycle 24 has peaked and will last the next two to three decades in a Dalton-class type hibernation like that of 1793 to 1830 in accordance with the RC theory. From this chart, we see the Little Ice Age was made up of three complete and separate 206-year cycles of the Sun.

The carbon-14 isotope readings are used in this case to reconstruct an analogous past global temperature curve. The scale on the left is gradations of C14 instead of temperature per se. Again, it is a proxy for temperatures going back far beyond the invention of the thermometer.

Few cases reach the strength of the evidence for a major earthquake anywhere in the USA more than the New Madrid Seismic Zone (NMSZ). Like clockwork, the NMSZ has produced region-wide, devastating earthquake or series of earthquakes with every 206-year solar hibernation or significant decline in the Sun's energy output since the year 1450!

The fact that another solar hibernation has begun should put every state in the area and the federal government on high alert. Our leaders should be making immediate preparations to withstand the coming great quakes with a goal of minimizing property damage and loss of life. If we have another solar hibernation predicted to be similar to the Dalton minimum, then there is a greater-than-80% probability the NMSZ will have another catastrophic earthquake or series of earthquakes between 2017 and 2038.

As shown in chapter 1, state and federal authorities have had numerous warnings of the threat facing residents in and around the NMSZ. Sadly, this warning has been ignored by the states and the federal government. Other than voluntary, token, "duck, cover and grab hold" exercises done in some schools and some public institutions in what are called the great shakeout drills, little is being done for what is a high probability of becoming America's worst ever geophysical disaster. The extended U.S.-wide effects of this predicted event will be far ranging and may be equal to or greater than that predicted for a rupture of the entire length of the Cascadia Subduction Zone or an M8.0 in San Francisco Bay or in the greater Los Angeles basin.

The unbelievable fact is that based on publically available information, it is clear that the U.S. government, including DHS/FEMA and the DOI/USGS, have effectively ignored our warnings of the looming geophysical catastrophes coming to the USA even though they have been substantiated by the solid science provided to them.

So what happened during the last solar hibernation in the NMSZ—like the one that is now predicted for the next two decades—that we should be so concerned about? Can we be better prepared for the next NMSZ event by learning what happened the last time? Yes, we can. Here are some of the accounts of the incredible series of earthquakes that struck the central Mississippi valley from December 1811 to February 1812:

It was a colder-than-normal winter in December 1811 in the central United States just south of the junction of the Mississippi River and the Ohio River along the border between Arkansas and Tennessee when the world was thrown into chaos. A great series of earthquakes struck along the Mississippi River, ending with a final catastrophic temblor near the small wilderness community of New Madrid. The land heaved and shook violently. Perhaps four mighty quakes ranging from M7.0 to M8.0 struck between December 1811 and February 1812. At one point, the great Mississippi River was thrust upward and flowed backwards for a brief period of time. What used to be the course of the river was cut off permanently and today is called Reelfoot Lake. A large number of fissures tore through the surface of the land for many miles.

Here is what the USGS describes just some of the incredible land deformations and areas where the land simply sank (subsidence) during the 1811–1812 New Madrid earthquakes:

> A notable area of subsidence that formed during the February 7, 1812, earthquake is Reelfoot Lake in Tennessee, just east of Tiptonville dome on the down dropped side of the Reelfoot scarp. Subsidence there ranged from 1.5 to 6 meters, although larger amounts were reported.
>
> Other areas subsided by as much as 5 meters, although 1.5 to 2.5 meters was more common. Lake St. Francis, in eastern Arkansas, which was formed by subsidence during both prehistoric and the 1811–1812 earthquakes, is 64 kilometers long by 1 kilometer wide. Coal and sand were ejected from fissures in the swamp land adjacent to the St. Francis River, and the water level is reported to have risen there by 8 to 9 meters.

The full USGS description of the NMSZ 1811–1812 CGE is shown at appendix 2.

These and other accounts in this chapter describe the largest series of quakes in U.S. history and what incredible force Mother Nature can deliver.

What has come to be called the New Madrid fault or the New Madrid Seismic Zone (NMSZ) still resides along the Mississippi River at the juncture of Missouri, Kentucky, Tennessee, and Arkansas, along with its smaller cousin just to the north, the Wabash Seismic Zone. See the map below:

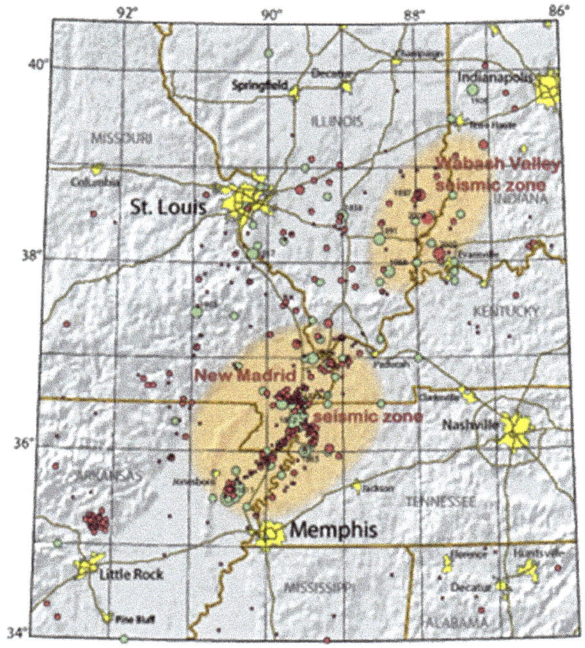

Figure 13. The central Mississippi valley with the two major fault zones, the New Madrid Seismic Zone and the Wabash Zone. This map shows earthquakes (circles) of the New Madrid and Wabash Valley seismic zones (orange patches). Red circles indicate earthquakes that occurred from 1974 to 2002 with magnitudes larger than 2.5 located using modern instruments (University of Memphis). Green circles denote earthquakes that occurred prior to 1974 (USGS Professional Paper 1527). Larger earthquakes are represented by larger circles. Source: USGS.http://earthquake.usgs.gov/earthquakes/states/events/1811-1812.php

Research by the authors examining the threat posed by the NMSZ is found in appendix 2. It contains the full research paper published in the Global Climate Status Report by the SSRC. The paper is titled "New Madrid Seismic Zone, central USA: The great 1811–12 earthquakes, their relationship to solar cycles, and tectonic settings."

The paper describes the history and geology of the region, including the major quakes that have struck there since 1450 and in particular the great 1811–12 earthquakes, along with their relationship to solar cycles and tectonics.

The long history of the NMSZ reinforces how active this region is and why it has a bad habit of delivering not just one catastrophic quake but a series of them. From the USGS NMSZ history (appendix 2):

"The geologic record of pre-1811 earthquakes reveals that the New Madrid seismic zone has repeatedly produced sequences of major earthquakes, including several of magnitude 7 to 8, over the past 4,500 years."

In this detailed technical paper by this book's authors, the tectonic layers of the planet around the NMSZ are exposed as is the history for major quakes in the region. Reproduced in part here in this chapter is the less-technical essence of that paper and its clear warning of the potential for another devastating earthquake in the NMSZ in the near future. We have recently created a new chart that takes the data from that paper and depicts the vital relationship to past climate changes.

There are several estimates of the magnitude and number of historic earthquakes that struck the NMSZ during the winter of 1811–1812. According to the USGS records, there were three main shocks, M7.5, 7.3, and 7.5, on December 16, 1811, January 23, 1812, and February 7, 1812, respectively, with a major aftershock of M7.0 on the first day (http://earthquake.usgs.gov/earthquakes/states/events/1811-1812.php). Other researchers, recorded in the source paper for the figure above, cite Nuttli (1987) who listed six M7.0+ quakes that include two M8.0+ earthquakes. Of them, two largest quakes were considered the greatest earthquakes in continental North America (Johnston and Schweig, 1996).

It is therefore reasonable to conclude there were at least three major quakes with a possible range from M7.3 to M8.0. In any case, "The series of catastrophic earthquakes that struck the central Mississippi River valley region from December 1811 to February 1812 were the most powerful series of earthquakes ever recorded in the North American continent. In terms of the shear force and energy release potential, this area of the United States remains to this day, the most dangerous earthquake zone in the USA with a capability of delivering a catastrophic series of earthquakes at any time. Our analysis suggests that we are on the precipice of yet another highly destructive series of calamitous earthquakes in the NMSZ."

When the NMSZ erupted in 1811–1812, the United States of America was a young nation with only a little over 5,300,000 citizens, not including Native Americans.[5] Back then, the NMSZ was still a wilderness with only a few trading posts and small towns. The first steamboat was in fact making its maiden voyage when disaster hit. Accounts from those in the area only partially describe what happened.

From the city of New Madrid, Missouri, historical records we have this:

"After the February 7 earthquake, boatmen reported that the Mississippi actually ran backwards for several hours. The force of the land upheaval 15 miles south of New Madrid created Reelfoot Lake, drowned the inhabitants of an Indian village; turned the river against itself to flow backwards; devastated thousands of acres of virgin forest; and created two temporary waterfalls in the Mississippi. Boatmen on flatboats actually survived this experience and lived to tell the tale."

And it was not just a series of several large quakes that tore through the region. Also from the city of New Madrid:

"As the general area experienced more than 2,000 earthquakes in five months, people discovered that most of crevices opening up during an earthquake ran from north to south, and when the earth began moving, they would chop down trees in an east- west direction and hold on using the tree as a bridge. There were 'missing people' who were most likely swallowed up by the earth. Some earthquake fissures were as long as five miles."

The area from Arkansas's northeastern corner to New Madrid was affected by sand boils or sand blows. These are geysers of sand caused by the liquefaction of the soil during the quakes that produced fountains and giant cones of sand that dotted the landscape. Remnants of these sand blows are still visible today. The longest was over 1.4 miles, but most were seen as large white patches in cultivated land where future crops failed to produce well or not at all for lack of nutrients. The following account, again from the city of New Madrid, discusses in some detail the events of the quakes and the first steamboat to traverse the Mississippi:

> The first steamboat travel on the Ohio and Mississippi Rivers took place during the New Madrid earthquakes. The New Orleans was out from Pittsburgh on October 20, 1811, bound for New Orleans. Captain Nicholas Roosevelt had brought along his young wife, their two year old daughter, and a Labrador dog. Ten days after leaving Pittsburgh, his wife Lydia gave birth to a son in Louisville, Kentucky. They waited a while for her to recover, and for the water to rise prior to crossing the

dangerous waters and coral reef at the Falls of the Ohio. On the night before the day of the earthquake, December 16, the steamboat was anchored near Owensboro, Kentucky, about 200 miles east of New Madrid, Missouri. Their dog, Tiger, insisted on staying in the cabin with them instead of sleeping on the deck.

Without realizing it, they were heading straight towards the epicenter of the greatest earthquake in American history. Their steamboat, intended to be an advertisement for steam travel, was thought instead to be the cause of the earthquake by many who saw it. At Henderson, Kentucky, where no chimneys were left standing, they stopped to visit their friends, the painter John James Audubon and his wife Lucy. Floating in the middle of the Ohio River they were protected from the earthquake tremors shaking the land, but not from the hazards of falling trees, disappearing islands, and collapsing river banks. After entering Indian Territory on December 18th, they were chased by Indians who figured the "fire canoe" had caused the earthquake, but they managed to escape capture by outrunning them. They even had a small cabin fire that night which they managed to put out.

Thousands of trees were floating on the waters of the Mississippi as they approached New Madrid on December 19th, three days after the earthquake. They found that the town of New Madrid had been destroyed. They didn't dare to stop and pick up a few survivors, for fear of being overrun, and they were without supplies. Most alarming was the fact that they had not seen a boat ascending the river in three days. They saw wrecked and abandoned boats. It was undoubtedly a miracle that they survived and kept on going. They tied up at one island, and the island sank during the night. Their dog, Tiger, alerted them to oncoming tremors. On December 22, they encountered the British naturalist John Bradbury on a boat at the mouth of the St. Francis River, who told them the town of Big Prairie was gone.

They arrived at Natchez, Mississippi on December 30 and celebrated the first marriage aboard a steamboat on December 31st, when the steamboat engineer married Lydia's maid. They arrived at New Orleans on January 10, 1812, safe and sound, after traveling 1,900 miles from Pittsburgh on the first steamboat to travel the western waters.

Figure 14. A nineteenth-century print of New Madrid earthquake chaos. Source: http://thumbs.media.smithsonianmag.com//filer/Phenomena-earthquake-central-631.jpg__800x600_q85_crop.jpg. Original image from the *Smithsonian Magazine* (Granger Collection, NYC)

If the NMSZ Had a Mega-Quake Series Today

Should the same or similar series of devastating quakes strike today with millions of Americans and large modern cities in the area, the destruction would be far and away the worst ever natural tragedy our country has faced. It would not be just the direct destruction to property and loss of life that would come from the quakes themselves but also the indirect and collateral damage that would come from the damage happening during the modern era with its sophisticated transportation, communications, social, and energy infrastructure. Here are just some of the major areas of damage that were not of consideration 200 years ago:

1. All the modern cities and towns that are located in the eight-state NMSZ are connected to and interconnected by power grids that provide heating and cooling to homes and businesses during the annual change of four seasons. As we already know, the last

NMSZ event happened midwinter 1811–1812. Imagine an eight-state loss of power that could possibly last for months during either the bitter cold of winter or, for that matter, the summer heat.

1. Cascading power grid losses may affect other states outside the NMSZ states. Should the power grids serving the NMSZ states fail during M7.5–M8.0 quakes, would they take with them neighboring state grids and widen the total power system losses far beyond these states? Energy companies and U.S. government energy agencies need to prepare for this scenario.

2. Fortunately, nuclear power plants have not been built in the NMSZ itself. However, there is a ring of twelve just outside it. Are these plants built to withstand a series of high M7.0 quakes or even M8.0 quakes? Are the backup power systems also able to take just huge jolts and keep operating so as to avoid a Fukushima-style catastrophe? Much has been learned from the March 11, 2011 Great Tohoku earthquake and tsunami that produced an M9.0 undersea quake and a destructive tsunami that killed at least 15, 984-plus people and heavily damaged many of the towns and villages along Japan's northeast coast.[6] The subsequent shutdown of the Fukushima nuclear reactor generators by the tsunami, that supplied power to keep reactor water cooled, failed. Three of four reactors then overheated and went into meltdowns, leading to a disastrous release of radiation. Contamination in the area, including the water table and surrounding land, continues to this day, and many residents are still not permitted nor desire to return to their irradiated homes.

Are we at risk of nuclear reactors around the NMSZ from failing? The graphic below shows the location of existing U.S. and central Mississippi nuclear reactors.

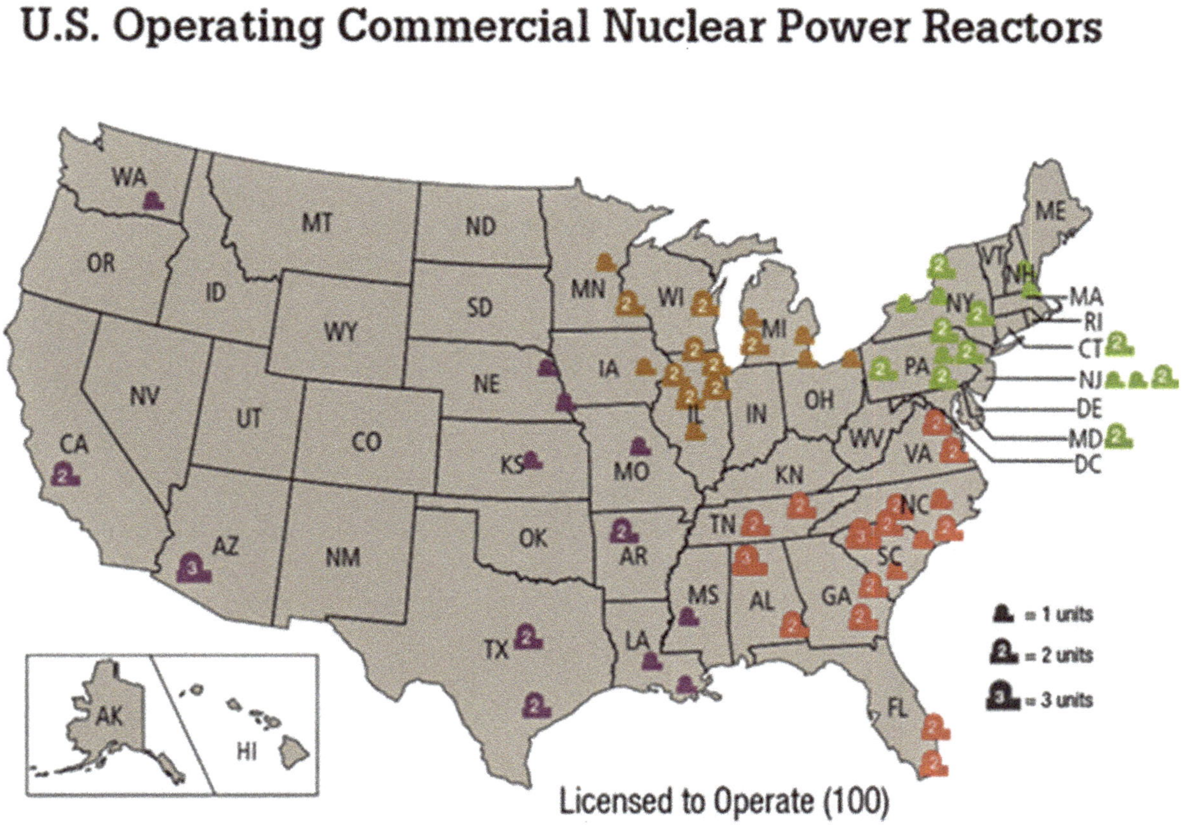

Figure 15. Nuclear reactors across the USA. Source: Nuclear Regulatory Commission (NRC) as of November 2015. Note the dozen or so reactors around the NMSZ.

From this chart, we see that there are no nuclear reactors directly on top of the NMSZ, from where the Ohio River and the Mississippi River join down the river to Memphis, Tennessee. However, the loss of any one of the surrounding reactors to a Fukushima-style event would create serious ill effects for the central USA and states that would be downwind should a meltdown occur.

Are these reactors able to handle the predicted M7.5 to M8.0 quakes along with their backup systems? Remember, their backup systems may be dependent on oil and gas supplies themselves during a long period of time when such fuels may be unavailable.

3. Major ground transportation systems supplying the US economy with goods and services go through the NMSZ and across the Mississippi River. Once these state and interstate roadways and bridges are destroyed, they would have significant impacts on U.S. and NMSZ state commerce. Their loss, likely for months, would have a significant impact on the U.S. economy and national gross domestic product (GDP) output. Thus, most U.S. states would be affected. It is difficult to grasp the significance of the loss some portion of every major interstate in an eight-state region in the heart of our country. Disruption of supplies travelling across our country would be unprecedented and extensive. Rerouting of traffic around the affected states' destroyed bridges and interstates would clog already-burdened roadways that border the eight NMSZ states. Traffic jams on the U.S. interstate systems from the Gulf of Mexico to the Great Lakes would occur over night and last for many months.

4. Domestic and international air traffic would be adversely affected through loss of airport infrastructure, including runways that are broken apart and could no longer support either takeoffs or landings. In a national emergency of the scale we are expecting, this means that the primary method of delivery of emergency aid via air transport would be unavailable.

5. A major disruption of Mississippi River traffic is likely during another catastrophic NMSZ event. Today the Mississippi is a major route for transportation of goods and raw materials via barges. If we see another upturning of the Mississippi as before, that transportation route may be both unusable if not dangerous for the barges and crews that operate them.

6. Significant damage to the NMSZ states and the northeastern U.S. states through loss of critical oil and gas supply lines will add to the widespread collateral damage from the next NMSZ mega-quake.

Most Americans are unaware that should the NMSZ have a series of quakes similar to the 1811–1812 temblors during the last solar hibernation, the majority of the gas and much of the oil that is delivered to the northeastern states will be shut off, possibly for many months. The huge pipelines that go directly through the central Mississippi valley region will likely be completely destroyed by the predicted devastating quakes resulting in a substantial and immediate power and heating oil crisis in the

U.S. northeastern states along with major economic damage. Power plants in the northeast that depend on the oil and gas from Texas and Louisiana refineries would be cut off, adding more misery to a much broader area than for just the eight NMSZ states. As already discussed, there is a real possibility these disasters will strike midwinter as they have in the past, adding to the already-predicted extremely dangerous and deadly years ahead.

Here is the gas pipeline chart for the United States and the eight-state NMSZ region:

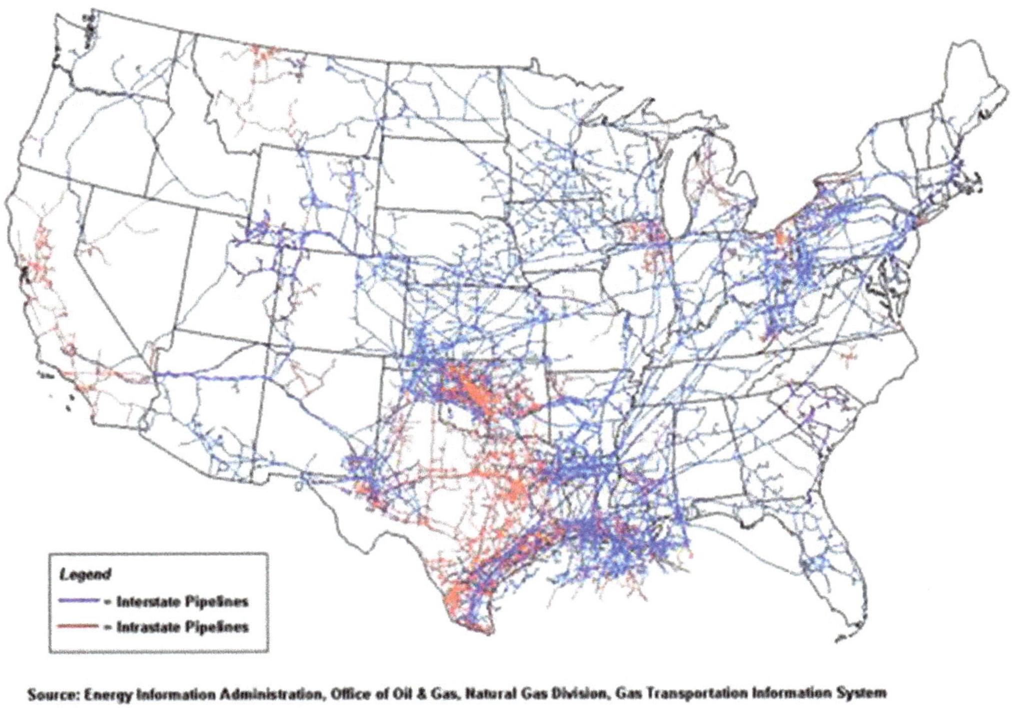

Figure 16. Natural gas pipeline network for the United States. Note the dark blue interstate gas pipelines that pass directly through the NMSZ. Source: Energy Information Administration (http://www.eia.gov/maps/)

This chart displays the major interstate and intrastate gas pipelines in the continental USA. It is clear from this chart that much of the oil and gas serving the Atlantic and northeastern states goes through one or more of the NMSZ states. The actual flow amount is such that the majority of the flow to the northeast goes through the NMSZ pipelines.

The damage from an NMSZ mega-quake has been studied by the U.S. government. Here are some of the FEMA estimates should such an event unfold today based on a 2009 study commissioned by FEMA:[7]

> The results indicate that Tennessee, Arkansas, and Missouri are most severely impacted. Illinois and Kentucky are also impacted, though not as severely as the previous three states. Nearly 715,000 buildings are damaged in the eight-state study region. About 42,000 search and rescue personnel working in 1,500 teams are required to respond to the earthquakes. Damage to critical infrastructure (essential facilities, transportation and utility lifelines) is substantial in the 140 impacted counties near the rupture zone, including 3,500 damaged bridges and nearly 425,000 breaks and leaks to both local and interstate pipelines. Approximately 2.6 million households are without power after the earthquake. Nearly 86,000 injuries and fatalities result from damage to infrastructure. Nearly 130 hospitals are damaged and most are located in the impacted counties near the rupture zone. There is extensive damage and substantial travel delays in both Memphis, Tennessee, and St. Louis, Missouri, thus hampering search and rescue as well as evacuation. Moreover roughly 15 major bridges are unusable. Three days after the earthquake, 7.2 million people are still displaced and 2 million people seek temporary shelter. Direct economic losses for the eight states total nearly $300 billion, while indirect losses may be at least twice this amount.

Twice that amount—$600 billion in assessable damages following another NMSZ rupture is the larger picture according to FEMA. That is ten times the $60.4 billion in U.S. government legislative funding approved for Hurricane Sandy that hit the Northeast on October 29, 2012.[8]

It is difficult for any of us to get our arms around the scale of devastation FEMA describes. What amounts to ten Hurricane Sandy's is striking at the same time!

But that is not the final estimate. The model used in the FEMA study allowed only one M7.7

earthquake, not three or four. Though they permitted the single quake to erupt along the entire NMSZ, the outcome from multiple catastrophic quakes surely leads to a different set of outcomes. That is certainly the case when one examines the need for immediate aid to those affected and reconstruction of the infrastructure—a process that would be quite difficult, if not impossible, under a multiple quake scenario.

What is routinely omitted from government studies and media articles of the impacts of another devastating NMSZ rupture is that again, we are looking at the prospect for multiple catastrophic quakes over several months. That means that those who go into the damaged areas to provide aid may themselves become casualties from additional major earthquakes. It also means that rebuilding would be impossible until the major quakes have passed to include the numerous aftershocks that could easily be above the M6.0 catastrophic level. Major relief work and reconstruction of roads, buildings, bridges, etc., may not be possible for as much as six months to a year after the first quakes begin.

****

Now that a new solar hibernation has begun, as they say, "It doesn't take a rocket scientist" to know whether another catastrophic quake is coming. The now widely disseminated chart at the beginning of this chapter showing the history of major NMSZ earthquakes during the last four solar hibernations or deep solar minimums was created to help people in the NMSZ understand what they are facing. This chart was also supplied to FEMA and the USGS along with notifications to all applicable state governors to serve as definitive evidence of impending quakes of record destructive capacity that are about to hit the New Madrid fault.

****

With the calculated bottom of the next solar hibernation as the year 2031, and measuring the difference between previous coldest years during solar hibernations to major quake and volcano occurrences, we have derived the window of highest risk period for geophysical catastrophes for the USA (and the planet) as between 2017 and 2038.

As the chart on solar hibernations and NMSZ quake occurrences shows, everyone in that multistate quake zone should immediately begin to make preparations for yet another catastrophic earthquake or series of quakes in the range of M6.8 to M8.0.

While a quake of the M6.8 magnitude like that of 1895 would be a blessing to the residents there as compared to a series of M7.3 to M8.0 temblors, even at the low end of the magnitude range, the destruction and loss of life would still be extensive. But what if we do see a repeat of the quakes of the 1811–1812 event? An M7.8 quake would be 32 times more powerful than the 1895 quake!

Unfortunately, even with this demonstrated risk, we still see no action from the U.S. government to prepare our country or the eight NMSZ states to a degree that is consistent with the level of the threat.

Getting across the seriousness of the situation can be aided by creating an analogous scenario. This following hypothetical story helps to bring the matter to the forefront of our ability to grasp how important the issue is:

Let's say you have read a recent newspaper article by an investigative journalist about a particularly dangerous intersection in your hometown. The journalist has searched police records and discovered for the past four years in a row that a fatal traffic accident has occurred at the same intersection in the same first week in December. Further, it seems the cause of the accident is always a runaway dump truck loaded with gravel that loses its brakes coming down the steep hill on one of the main roads leading into the intersection.

You dismiss the story as merely an oddity of nature and chance. Yet the next day, you find yourself and your family in your car heading toward that very same intersection. Once again, it is the first week in December, and as you near the intersection, you see a dump truck overflowing with gravel spilling over onto the roadway as it careens wildly, barreling down the hill. Is history repeating?

This time, the question has even more importance as the driver of the dump truck starts waving his hand wildly out his window and begins to flash his truck lights and sits on his horn as the dump truck picks up speed. Some pedestrians at the intersection start waving their arms and shouting to the traffic to not enter the intersection.

What will you do? Will you ignore the history of the last four years and what your eyes are telling you—pretending the dump truck will not cause another fatal accident, this time involving you and your family? Since you have the green light, you are authorized to keep driving through the intersection, but is it safe to cross? What do you, the driver, do?

The prudent and safety-minded person will of course protect his family and slow down to a stop as he approaches the intersection, making certain his car and the runaway dump truck do not collide.

There are several direct parallels between this analogy and the coming solar hibernation. Just like the dump truck causing four accidents in a row, we have had four catastrophic earthquakes in a row in the NMSZ, linked with four solar hibernations in a row. The earthquakes are of course the fatal accidents, the vision of the runaway dump truck hitting your car is the next solar hibernation, predicted to be just ahead as we can see from the declining sunspot records. The frantic pedestrians trying to warn others symbolizes the authors of this book and a host of other scientists as we try to warn everyone that catastrophic earthquakes are coming again. The newspaper story (i.e., history) is about to repeat, and it is now beyond our ability to control it!

Metaphorically, the driver of the car can also be viewed as not one of us—we the people. It is our elected leaders charged with our protection. We the people are the passengers in the car who, because we have no control over the steering wheel, have effectively ceded all our rights to our own protection to the driver of the car or, rather, our state and federal government and their scientific agencies.

As passengers seeing the dump truck and its fearful driver headed in our direction, do we choose to sit quietly while the car driver (our government) drives in front of the dump truck killing us all, or do we scream and yell to the driver to stop the car and not drive into the dangerous intersection?

If the driver of the car has blinders on, representing the U.S. government's refusal to accept that earthquakes can be predicted to any degree, and its failure to use proven climate science as its guide, then he will drive through the intersection without ever seeing the runaway dump truck. The driver and we the passengers will all perish. However, if the driver is observant of all that goes on around him, learns from history, and has good peripheral vision (i.e., has his eyes open to new research that shows that some earthquakes and climate variations are predictable), then he will slam on the brakes, and we passengers will be saved.

This then is the choice we all are now facing in the heart of our great interconnected country. Will we ignore the warning signs and the long history of the New Madrid Seismic Zone and its intimate relationship with the Sun? Will we live day to day with blinders on as the USGS, FEMA, and the rest of the federal government are doing as they intentionally disregarding the historical warning signs of the coming geophysical devastation simply because it does not fit the politically driven narrative of man-made global warming?

Will our leaders in these states and the federal government roll the dice on the lives of our people in the central Mississippi River Valley?

Will they continue to ignore or cover up the hard science that they know says that record earthquakes are about to strike the region to over 80% certainty beginning as soon as next year, 2017?

Chapter 4

The West Coast: Widespread Damage

Hope for a miracle. But don't depend on one.

—The Talmud

Dr. Thomas Jordan, director of the Southern California Earthquake Center, was correct when he said that the San Andreas Fault was "locked, loaded, and ready to roll" in an *LA Times* article published May 4, 2016. He was expressing a genuine concern like so many of us who study why and when earthquakes happen. It has been 160 years since the last big one hit the San Andreas. That was the M7.9 Fort Tejon quake of 1857, the largest ever in California's recorded history. It is easy to say that the next big one is overdue. Like the quote from the Talmud, Dr. Jordan may be hoping for a miracle that the San Andreas may take another 160 year breather, but he is not depending on it. Good sense, like the quote, may be the source of his efforts and that of many others in California to deploy the ShakeAlert earthquake warning system.

The ShakeAlert system may provide only 30 seconds' to 1 minute's warning for distant locales from the epicenter of a San Andreas rupture, but that's after the quake has already begun. Some have argued that this is a questionable amount of reaction time to shut down critical emergency and electrical power systems before they are damaged. However, the fact that the ShakeAlert system is struggling to get funding from the current administration and the state of California shows once more how incompetent governments are in making important decisions vital to public safety.

However, to assume that only the Los Angeles basin or the San Francisco Bay area are the only areas at risk with regard to West Coast is to seriously diminish the threat that we believe exists for the next decade and a half. Is it time for the big one to finally do its thing? Will the next big quake hit the San Francisco end of the San Andreas or the Los Angeles end, both at the same time, or will we see a zipperlike quake that starts at one end and runs the length of the fault? And what about the Cascadia Subduction Zone (CSZ)? We will provide our answers to some of these questions in this chapter.

****

The coldest period of the Little Ice Age (LIA) cold era (1350–1830) was the Maunder minimum (1615–1745). Near the center of the Maunder minimum on January 26, 1700, the Cascadia Subduction zone ruptured, creating the largest earthquake and tsunami the continental United States has ever seen estimated at M9.0. Only the great 1964 Alaskan M9.2 quake was larger. That quake sent a tsunami across the Pacific Ocean where it did damage but caused no major loss of life in the coastal villages as it came ashore in Japan in the early morning hours before anyone was up. The question then is this.

After 316 years of subduction of the ancient Juan de Fuca plate under the North American plate, is it time for the Cascadia zone to bring forth another megathrust quake and tsunami? What will this mean for the California-long San Andreas Fault that lies at the southernmost portion of the subduction zone? Some scientists within the IEVPC believe that other forces may be at work and that the subduction process is debatable. Regardless, and going with the consensus view of subduction for the moment, we maintain that the analysis of the West Coast threat using the RC theory may answer our central questions of when and where the next major catastrophic earthquakes will strike the region.

Figure 17. Cypress Viaduct in Oakland, California, after the 1989 M6.9 Loma Prieta earthquake. Source: USGS

The State of California Earthquake History

Without question, the state of California ranks at the top of America's conscience when the matter of earthquakes comes up. The state has the best-known quake history, and a majority of Americans have doubtless heard of the San Andreas Fault and can tell in what state it resides. Numerous TV stories and large-screen movies have dealt with storylines around a devastating California earthquake over the decades. There are plenty of actual stories throughout the state to draw on when it comes to large destructive earthquakes.

Aside from Alaska, which ranks at the top when it comes to the largest earthquakes over the past century, and the NMSZ with the largest series of US earthquakes, California is the U.S. state where most of our moderate to large quakes have struck since our country's founding. Here is how Wikipedia summarizes the West Coast earthquake environment:

California has numerous active faults throughout the state which are known to produce large earthquakes. The most active of these is the San Jacinto Fault Zone in Southern California, which has produced large events on a regular basis throughout recent history. The Mendocino Triple Junction located offshore of Northern California is also very active, producing many earthquakes above M6 throughout history. Northern California is also subject to megathrust earthquakes on the Cascadia subduction zone (extending north from Mendocino), such as the 1700 Cascadia earthquake, magnitude of approximately 9. The town of Parkfield in central California is located on a section of the San Andreas Fault that produces an earthquake of about M6 every 20–30 years on average in 1857, 1881, 1901, 1922, 1934, 1966 and 2004.

The largest recorded earthquake in California was the 1857 Fort Tejon earthquake, with an estimated magnitude of 7.9. This earthquake ruptured the San Andreas Fault from Parkfield to Wrightwood, a distance of 225 miles (350 km). The most destructive earthquake to date was the 7.8 magnitude 1906 San Francisco earthquake, when more than 3,000 people perished in the earthquake and the fires that followed. The 1906 quake ruptured the northern segment of the San Andreas Fault for 296 miles (477 km), from San Juan Bautista to near Cape Mendocino in the north. More recently, the 1989 Loma Prieta earthquake, which registered 6.9 and affected the San Francisco Bay Area, and the 1994 Northridge earthquake which registered 6.7 and hit the Greater Los Angeles Area, caused widespread damage and deaths in their respective regions.

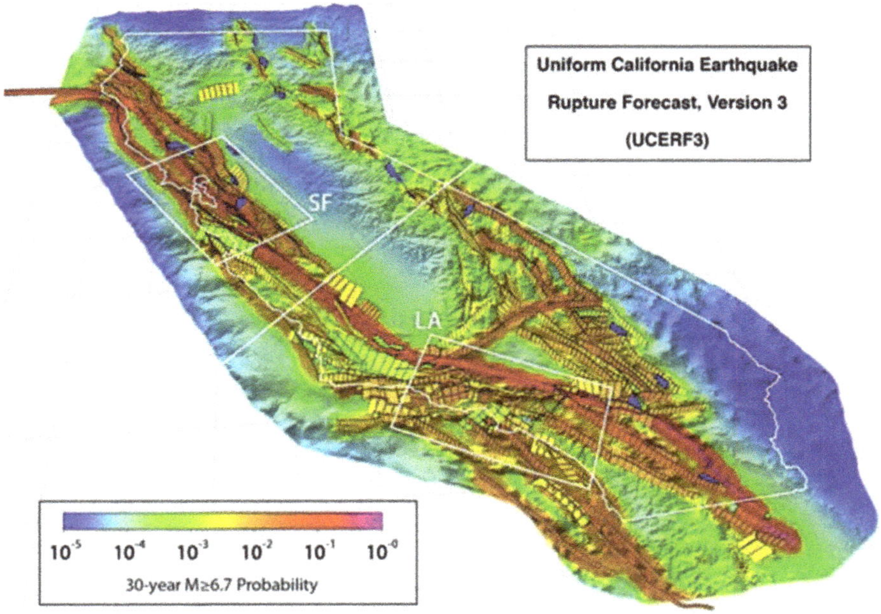

Figure 18. California Earthquake Threat. Three-dimensional perspective view of the likelihood that each region of California will experience a magnitude 6.7 or larger (M≥6.7) earthquake in the next 30 years (M6.7 matches the magnitude of the 1994 Northridge earthquake, and 30 years is the typical duration of a homeowner mortgage). Source: USGS.

The state's earthquake hazard map was recently updated in a December 2015 study titled the "UCERF3, Third Uniform Earthquake Rupture Forecast."[9] Produced by the USGS, the study clearly displays a state that is at high risk of major earthquakes. The chart above provides a substantially updated view of the California earthquake threat. The full report is referenced for this chapter in appendix 4.

In another excellent historical piece extracted from the UCERF3, we have the following chart showing when and where past California quakes struck.

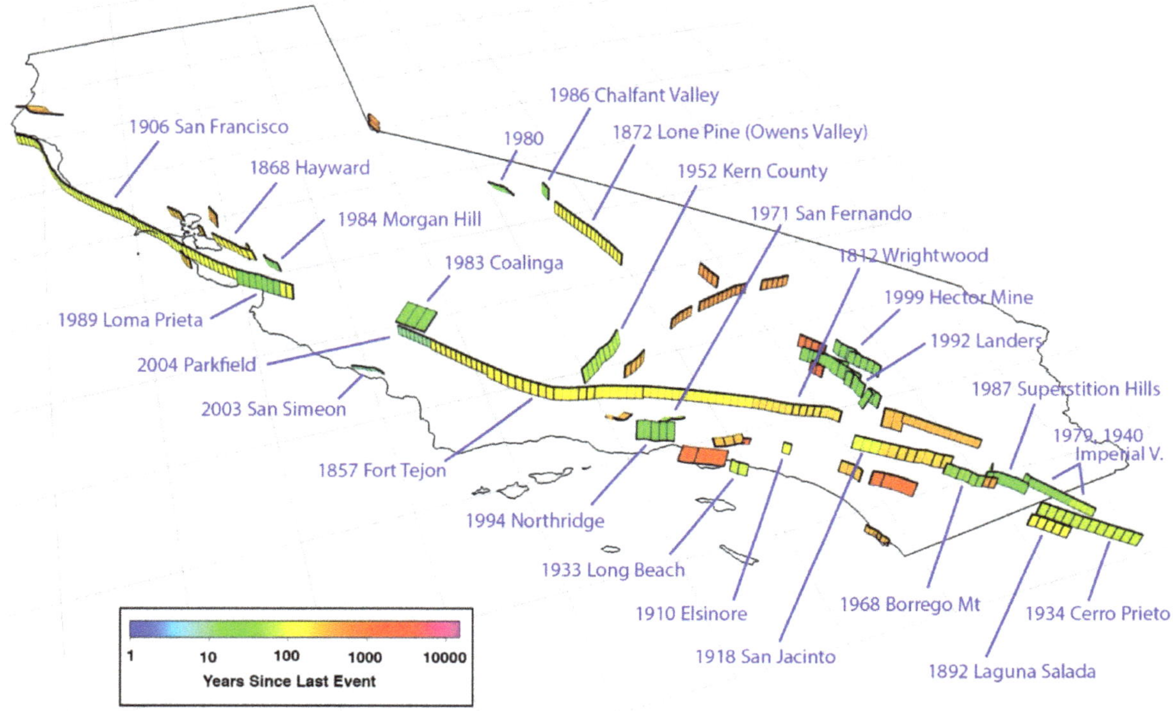

Figure 19. Historical chart of past California earthquakes. Source: USGS.

The chart above assists the USGS in explaining a concept that the likelihood of the next major quake is to be found where the longest time has passed. There are several points of departure between the USGS/federal government view of predicting earthquakes in California and what we are saying in this text:

1. As we might expect, the USGS does not link any earthquake risk to cold climate or solar hibernation events. It is politically unacceptable to discuss any climate related effects that are not tied to the myth that mankind controls the Earth's climate through emissions of CO_2.

2. We believe in earthquake prediction to the extent of improving the accuracy of the process, not that we or anyone else is yet near 100%. The government's position remains, tragically, that no one can predict earthquakes. Period.

3. The USGS position for earthquake probabilities across the USA is couched in carefully selected words and small percentages and falls far below the level of threat that our approach to earthquake analysis suggests.

The last catastrophic quake to strike the San Francisco Bay area was the great M7.8 quake of 1906. Once again, this quake occurred during a downturn in solar energy output, a function of what I called the centennial cycle in the RC theory. Does this mean that now that we are approaching the next iteration of the centennial cycle coincident with the bicentennial cycle and that San Francisco is about to be devastated again?

Using the onset of a solar hibernation as a guide, the answer to these questions is yes!

The probabilities are highest and the risk greatest for catastrophic earthquakes to strike during solar hibernations. Our research shows they are most likely to occur either just prior or when we reach the cold bottom of these solar cycles at the same time a new cold climate has arrived.

The Cascadia Subduction Zone

The widely described and planned-for next Cascadia Subduction Zone (CSZ) event is getting more attention. Thanks go in great part to numerous geologists and seismologists who have done research on the U.S. northwest coast for decades, and some recent media focus there. Perhaps more importantly, the March 11, 2011 Great Tohoku earthquake and tsunami that devastated the northeast coast of Japan in the Sendai province brought home just how real mega-quakes along the Pacific Rim area are. The Japanese M9.1 quake killed at least 15,600 and destroyed many coastal villages. Secondary effects caused the explosion of three of four nuclear reactors at the Fukushima Daiichi power station, leading to a nuclear radiation disaster. The resultant tsunami generated waves over 33 feet, much higher in a few locations, and spread across the Pacific Ocean, striking the northwest coast of the USA.

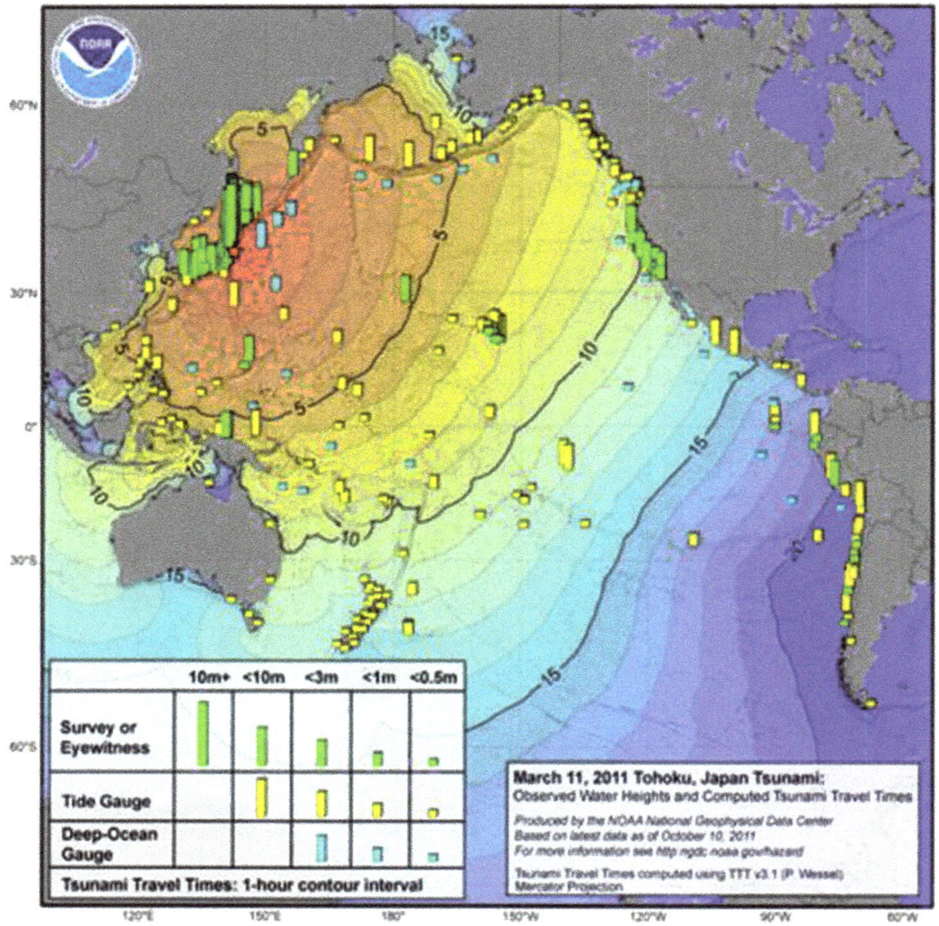

Figure 20. A chart depicting the spread of the Great Tohoku earthquake tsunami. Note the waves heights over 33 feet (10m) near Japan and waves approaching 3 meters (1 foot) on the other side of the Pacific as the tsunami hits the USA and Chile. Source: NOAA.

The existence of a distant past M9.0 megathrust earthquake and tsunami along the Oregon and Washington State coastlines has stirred the imagination of millions of Americans who live under this constant threat. The challenge for researchers and government officials has been how to anticipate and prepare for the next CSZ strike. We believe, like for most major earthquakes, that a study of how they coincide with solar hibernations may provide the answer to not only the when but how much of a rupture of the zone may occur for the next CSZ event.

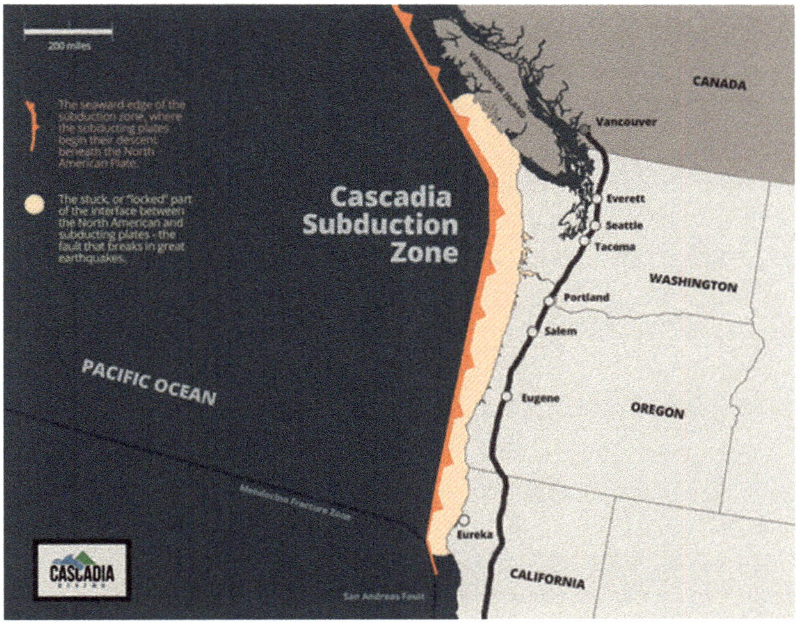

Figure 21. The Cascadia Subduction Zone. Showing boundaries for northern California, Oregon, Washington State, and Vancouver Island, Canada. Source: FEMA.

What most Americans are not aware of is that the history of the last CSZ was not disclosed until relatively recently when several researchers' individual works came together to reveal the complex story of the January 26, 1700 CGE. Here is a brief, though, informative accounting of this story from StFrancis.edu:

The Year 1700 Cascadia Earthquake

> On January 6, 1700, at about 9:00 p.m. Pacific Standard Time, a gigantic earthquake occurred sixty to seventy miles off the Pacific Northwest coast. The quake violently shook the ground for three to five minutes and was felt along the coastal interior of the Pacific Northwest. A tsunami formed, reaching about 33 feet high along the coast, and then traveled across the Pacific Ocean and hit the east coast of Japan.

> The earthquake ruptured what is known as the Cascadia subduction zone—the area of overlap between two of the tectonic plates that make up the earth's surface. These plates are the Juan de Fuca and the North American. The Cascadia subduction zone extends from Vancouver Island, British Columbia south to Cape Mendocino in Northern California. The earthquake dropped the entire Pacific Northwest ocean coastline three to six feet.
>
> The tsunami traveled across the Pacific Ocean for some 10 hours and at midnight on January 27, 1700, local time, it hit the east coast of Japan. By then the tsunami was only 6–10 feet high but it still did significant damage.
>
> The very existence of this magnitude 9 earthquake was unknown just twenty years ago. An early breakthrough came in 1987 when U.S. Geological Survey geologist Brian Atwater reported geological traces of giant earthquakes along Washington's Pacific coast. These traces included groves of trees that were killed when an earthquake lowered forests into the salt water. Another important clue was reported a few years later when the earthquake was dated to the decades between 1680 and 1720. This clue came from radiocarbon tests by Minze Stuiver of the University of Washington. Meanwhile in Japan, several researchers were following these developments. They consulted their nation's archives of old writings about earthquakes and tsunamis. For the period between 1680 and 1720 they found one orphan tsunami that could have come from Cascadia. That tsunami occurred in January of 1700. Finally, a final clue was found in the groves of killed trees on the Washington coast. Using annual growth rings in the trees, David Yamaguchi of Seattle and Gordon Jacoby of Columbia University showed that the trees lived through the 1699 growing season but were dead by the following spring—exactly the dates expected if the earthquake occurred in January 1700.

One of the most impressive pieces of scientific research done to date on the CSZ came in the form of a USGS study published July 17, 2012. This comprehensive multiyear study done by Oregon State University analyzed ocean floor sediment deposits (turbidites) in the Pacific Ocean off the California, Oregon, Washington, and western Canadian coasts (i.e., Vancouver Island). See the reference for the paper at appendix 7. The full 184-page study can be downloaded at no cost for those who are interested in the CSZ or just for those who want to see what a truly superior scientific study looks like.[10]

Among the rich set of findings in the study are several jewels that pertain to this book's correlation of solar activity and earthquakes. For example, the study found that the full CSZ has four rupture modes from Vancouver Island (Canada) down to Northern California. Interestingly, the four segments rupture at differing time frames with the most southern ruptures occurring more often and on shorter cycles than the northern segments of the CSZ. Segment distances shown are approximate, and are supplied by the authors in the table below:

Cascadia Subduction Zone Segments

Segment	Rupture Cycle, USGS Study
Segment A: the full CSZ, including the northern segment (rupture mode), the distance between Mendocino, CA through about the first 100 miles of Vancouver Island	every 500–530 years
Segment B: approximately 2/3 of the southern portion of the CSZ, the approx. distance between Mendocino, CA, and Cape Shoalwater, WA	every 410–500 years
Segment C: approximately 1/2 of the full CSZ in the southern portion, the approx. distance between Mendocino, CA and Newport, OR; this is close to an existing 354-year solar cycle.	every 300–380 years
Segment D: approximately 1/4 of the southern portion of the CSZ, the approx. distance between Mendocino, CA, and Coos Bay, OR; this is a close match to both the 206-year and a 230-year solar cycle.	every 220–240 years

Additional findings and conclusions are extracted from the study as follows:

From the study abstract:

> The average age of the oldest turbidite emplacement event in the 10–0-ka series is 9,800±~210 calyr B.P. and the youngest is 270±~120 calyr B.P., indistinguishable from

> the A.D. 1700 (250 calyr B.P.) Cascadia earthquake. The northern events define a great earthquake recurrence of ~500–530 years.
>
> The sequence of 41 events defines an average recurrence period for the southern Cascadia margin of ~240 years during the past 10 k.y.
>
> Time-independent probabilities for segmented ruptures range from 7–12 percent in 50 years for full or nearly full margin ruptures to ~21 percent in 50 years for a southern-margin rupture. Time-dependent probabilities are similar for northern margin events at ~7–12 percent and 37–42 percent in 50 years for the southern margin. Failure analysis suggests that by the year 2060, Cascadia will have exceeded ~27 percent of Holocene recurrence intervals for the northern margin and 85 percent of recurrence intervals for the southern margin.

From Page 133 of the study:

> Thus we suspect that the next event would most likely be in the southern Cascadia. The general pattern of at least one smaller event between the larger events holds true for 17 of the 19 northern margin intervals, suggesting a 90-percent chance the next event being a southern-margin event.

We agree. With the passage of the normal intervals already met for the segment D events without a rupture of the southern Cascadia segment (i.e., 220–240 years), we have advanced to the next time frame for a segment C event. With the last event having taken place in 1700, and a Segment C range of 300–380 years, a little math places the next rupture between the year 2000 and 2080. Given the estimate of 85% noted in the study for a southern segment rupture by 2060, and that 17 years have elapsed in the long-range forecast, we are now down to a time frame between 2017 and 2060, a 43-year spread.

The next question then becomes, can we identify an even smaller time frame, something we can hang our hat on and start planning for? The answer can be found with the already-established solar hibernation correlation. If we can find that the majority of Cascadia events strike during a downturn in solar energy output as in other earthquake series, then the next solar hibernation is our most likely time frame for another CSZ strike. But is this so? We are about to see.

One of the other findings from the OSU 2010 study was that the northern San Andreas quakes may be directly related to southern Cascadia events. The implication is that the Great San Francisco quake of 1906 may be a result of destabilization of the far most southern portion of the CSZ, just to the north of San Francisco. This is important. Demonstrating such a connection yields new information for preparation activities for those living within the high risk northern San Andreas as well as the southern Cascadia region. If the southern CSZ is due for a rupture then this may also trigger another San Francisco quake.

In the chart below, we have plotted C14 isotope readings from the INTCAL2004 analysis by Reimar et al. and a data set also used by the OSU study team headed by Dr. Goldfinger. We have identified past quake dates along the CSZ knowing the time of the last one (1700) and the starting time frame of the INTCAL 2004 C14 analysis which gives us a last dated C14 reading in 1950. Thus, the time since 1950 and 1700 is 250 years—a time frame that again puts us outside the segment D rupture period and places the next event in the segment C category.

This chart, like all solar-terrestrial analyses, produces much knowledge. Here we see that four of the last five Cascadia events (including the supposedly unrelated 1906 event) happened during a drop in solar activity if not a full-blown solar hibernation. The last 1,500 years were chosen for this assessment rather than the full ~10,000 years in the OSU study so as to show most recent CSZ trend activity.

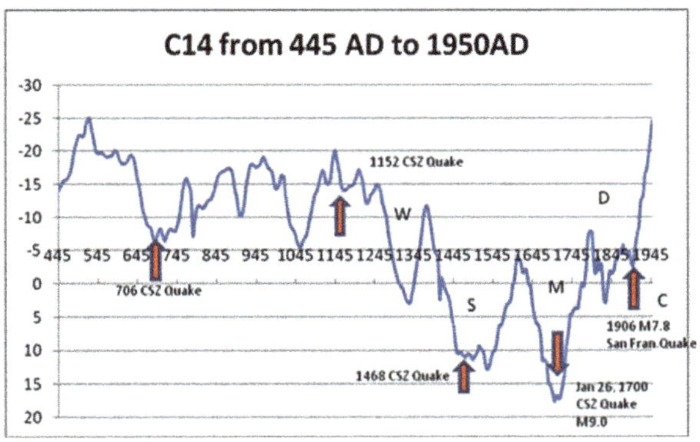

Figure 22. Full Cascadia Subduction Zone ruptures as compared to solar activity measured by C14. The reduced periods of solar activity are noted by the *W* for the Wolf minimum, *S* for the Sporer minimum, *M* for the Maunder minimum, *D* for the Dalton minimum, and *C* for the centennial minimum when the 1906 San Francisco quake struck, which is included for reference only and not as a CSZ rupture per

se. Source: J. Casey, C14 Data. INTCAL2004 Reimar, et al., Earthquake data from USGS Paper 1661-F, Goldfinger, et al., 2012.

Since the chart is inverted on the x-axis scale, we view these like global temperature charts which show "down" as being colder periods, i.e., Sporer and Maunder minimums. Note there was no CSZ quake during the Dalton minimum (1793–1830) or the Wolf minimum among others. What does this mean? It could signify that all or part of the energy that would normally have been expended in a far southern segment D quake during the Dalton minimum has been carried forward or, rather, building up for a larger segment C area rupture from Mendocino, California, to Newport, Oregon. Similar to the lower San Andreas, are we in a situation where the CSZ is also locked, loaded, and ready to roll? Dr. Choi and I think so and may involve the southward movement of energy along the main fault.

This analysis leads to an important consideration:

It is not necessary to show that a CSZ rupture happens during every solar hibernation or period of reducing solar activity. It is only necessary to show that when they do take place, it is during one of these reduced solar activity periods. For the last four CSZ eruptions dating back to 706 AD, all, 100%, took place during declining solar activity, and three of the four (75%) took place at the bottom of a solar hibernation.

Knowing when the next such reduced period of the Sun arrives adds greatly to our ability to predict the next CSZ, especially when coupled with the OSU study which shows which segment will produce a mega-quake during specific periods. In effect, the two approaches from differing fields of science, one cold-climate related and one turbidite-accumulation based, reinforce each other's conclusions. Sounds a bit like the big bang discovery doesn't it?

The study also cited the possibility that the most southerly portion of the CSZ at the intersection of the Mendocino fault might erupt every 200 years or so. Is this related to the 206-year solar cycle? The study cited the entire southern segment had on average a rupture every 240 years. It should also be remembered that the study rightfully indicated there was a large standard deviation for these averages. This means that ruptures of each segment of the CSZ could happen well before or well after the stated average years of each segment's cycle.

The study further noted that ruptures along the most southern segment may also trigger northern San Andreas Fault ruptures and cited the possibility that the 1906 San Francisco quake was related to a previous small rupture near the southernmost end of the CSZ. The northern portion of the San Andreas Fault zone begins near the Mendocio fault, i.e., the start of the southern CSZ. A quote from the study explains the implications for another major rupture in segment D: "It also suggests the possibility of two damaging earthquakes on the west coast, closely spaced in time."

This study, one of the most comprehensive and thoroughly conducted ever, is exciting to those of us who seek well-researched answers about what the history of the CSZ mega-quakes have been.

More recently, attention was focused on the area through an article that ran in *The New Yorker* by Kathryn Schultz on July 15, 2015. The author received a Pulitzer Prize for the well-written article and deservedly so. The most famous quote from the article that has made the rounds was that "anything west of I-5 would be toast" should another CSZ event happen, according to Kenneth Murphy a FEMA Region X leader.

In addition, within the OSU study, there are some important comments from the study authors:

From Dr. Goldfinger, study leader:

> Over the past 10,000 years, there have been 19 earthquakes that extended along most of the margin, stretching from southern Vancouver Island to the Oregon-California border . . . These would typically be of a magnitude from about 8.7 to 9.2—really huge earthquakes.
>
> We've also determined that there have been 22 additional earthquakes that involve just the southern end of the fault . . . We are assuming that these are slightly smaller – more like 8.0 – but not necessarily. They are still very large earthquakes that if they happened today could have a devastating impact.

Patrick Corcoran from the OSU Sea Grant Extension program had this summarization for the entire study:

> The research, though, is compelling. It clearly shows that our region has a long history of these events, and the single most important thing we can do is begin 'expecting' a mega-quake, then we can't help but start preparing for it.

Well said.

This is not the only independent analysis that suggests that the Cascadia Subduction Zone (CSZ) and the U.S. West Coast are on the verge of catastrophic earthquakes.

Work by Dr. Choi, and Dr. Tsunoda have pinned down the coming West Coast, and especially, Southern California earthquake threat. Their combined analysis of the flow of crustal/mantle energy over time identifies a period over the next few years to a decade or so, of shallow earthquakes of significant size. Extracted here are some of the findings and conclusions of their research, documented in part by their paper titled "Seismo-Volcanic Energy Propagation Trends in the Aleutians and North America." The full paper is in appendix 2.

Concluding remarks

> This study clarified the energy flow occurring in North America and the Aleutians by analyzing the time-space distribution of major earthquakes and volcanic eruptions. Throughout the region, a consistent earthquake energy propagation rate in the shallow mantle and crust was obtained; 120–140 m/day. This rate is slower than that of the western Pacific and South America.
>
> A complex flow pattern is observed in the Aleutian Islands—a mixture of eastward and westward flows: 1) Westward volcanic energy flow in the Aleutians is much faster than earthquakes, probably due to its deeper root, 100 to 250 km. Its speed is, 180–225 km/year or 500–620 m/day. 2) Eastward counter flow, which is slower than the westward flow, was also detected in the volcanic distribution pattern, whose average speed was calculated as 67 km/year or 185 m/day. The eastward flow of seismic energy is also observed from 1978 to 1990 in the eastern Aleutian earthquake distribution; it turned out to be almost same in speed as the volcanic energy flow. This eastward flowing energy obviously came from the west, through the Japan-Kuril-Kamchatka route.
>
> The source of energy in the study area is considered the South West Pacific. The energy transmigrates in the middle mantle channel through Hawaii, and upwells in the Gulf of Alaska, where the major portion of the energy bifurcates westward and southeastward along the coast of the Aleutian and North America.

The emergence of middle mantle energy to the shallow depth in the Gulf of Alaska is supported by the strong regional SST anomaly in the same region since December last year (2013). The discharged energy from this region should generate another cycle of strong earthquake and volcanic activities down the streams in the coming decades in the Aleutians and North America. Coupled with the deepening solar hibernation with the heightened endogenous energy release (Casey, 2012; Choi, 2013; Choi et al., 2014), it may bring a series of disastrous effects—catastrophic earthquakes and volcanic eruptions. Here, the knowledge of the upwelling site and propagation speed should help successfully predict the time and locality of these natural disasters.

The current work confirmed the veracity of energy migration inside the Earth through a complex network of flow channels developed mainly in the Circum-Pacific Meso-Cenozoic mobile belts. Further studies are needed to fully understand the energy flow phenomena; flow channels, their geological control, Earth rotation effect, internal workings and processes of the deep Earth, and interaction with other planetary forces.

Their analysis of the time-based arrival of earthquakes for the West Coast by latitude is show in the figure below:

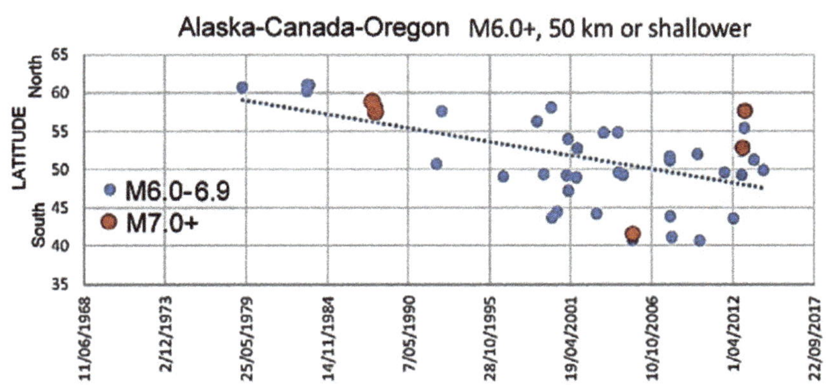

Figure 23. Chart of time-based earthquakes identified with deep Earth energy flow. This chart shows an increasing cluster of major earthquakes approaching 45 degrees north latitude by 2017. Source: D. Choi, F. Tsunoda.

From the chart above, we see that at latitude 45 degrees north, roughly that of Eugene, Oregon, the energy from deep within the Earth reaches the center of the Cascadia Subduction Zone entering segment C of the CSZ. This portion of the West Coast has already come within the high-risk period as identified by these scientists. We should expect this trend to continue with major to catastrophic earthquakes in this area at any time beginning in 2017 and thereafter!

This excellent piece of research finds the same growing threat to the U.S. West Coast and California that was also detected through sunspot analysis—yet another corroboration of the link between the Sun's behavior and its impacts on the Earth.

To restate the important conclusion this energy flow study by Dr. Choi and Dr. Tsunoda, covering much of the planet's surface, we have the following:

> The discharged energy from this region should generate another cycle of strong earthquake and volcanic activities down the streams in the coming decades in the Aleutians and North America. Coupled with the deepening solar hibernation with the heightened endogenous energy release (Casey, 2012; Choi, 2013; Choi et al., 2014), it may bring a series of disastrous effects—catastrophic earthquakes and volcanic eruptions.

Here then is what we have concluded regarding the possibility of a Cascadia Subduction Zone rupture in the near future:

1. Given the time interval from the last 1700 CSZ event, it is possible that the next rupture of the CSZ could be a full CSZ zone rupture around M9.0.

2. However, we believe the next CSZ rupture is more likely to be a lesser, though still quite destructive, M8.0 segment C rupture—the distance between Mendocino, California, and Newport, Oregon. Deep energy flows track to this area at this time.

3. The rupture time frame based on solar cycle analysis is between 2017 and 2033—the latter year being just a couple years after the predicted bottom of the next predicted solar hibernation in 2031. Note: the NMSZ time frame cited in chapter 4 was only slightly longer at 2017 to 2038.

4. We believe the current U.S. and state government risk assessments for the CSZ are dangerously conservative as to when the next CSZ rupture may occur. This may lead to complacency among the populace and hence lack of preparation for the next catastrophic CSZ event.

5. Assuming that no CSZ rupture takes place during the just-started solar hibernation, that would yield a three of five or 60% occurrence rate over past solar hibernations. That then is our lowest estimate for the next rupture in the CSZ—a minimum 60% probability within the next solar hibernation.

Now that we have looked at the CSZ as such, what remains is a similar assessment for the San Andreas and related California faults. For simplicity, we will look at the record of all California quakes available from the USGS since 1811, the time frame for which the USGS has reliable records. Coincidently, 1811 was the bottom of the last solar hibernation, the Dalton minimum.

In this case, we are dealing with a much shorter time frame, 230 calendar years versus the 1,500 years of the CSZ analysis. However, we can use more detailed sunspot records which began in 1610 to do this analysis. The sunspot trend is almost identical with global temperatures especially during times of low solar activity. We will see whether the cold climate relationship to earthquakes is present in California as it was in the NMSZ and along the CSZ.

It might be useful to understand the big picture of what the Earth's climate has been doing for the last 11,000 years to put into perspective the last two centuries of solar activity. We can do this by examining deep ice cores taken from Greenland. Thanks to the work done by Dr. Ole Humlum at Climate4you.com, we have a striking comparison of global temperatures over the millennia along with the change in atmospheric CO_2:

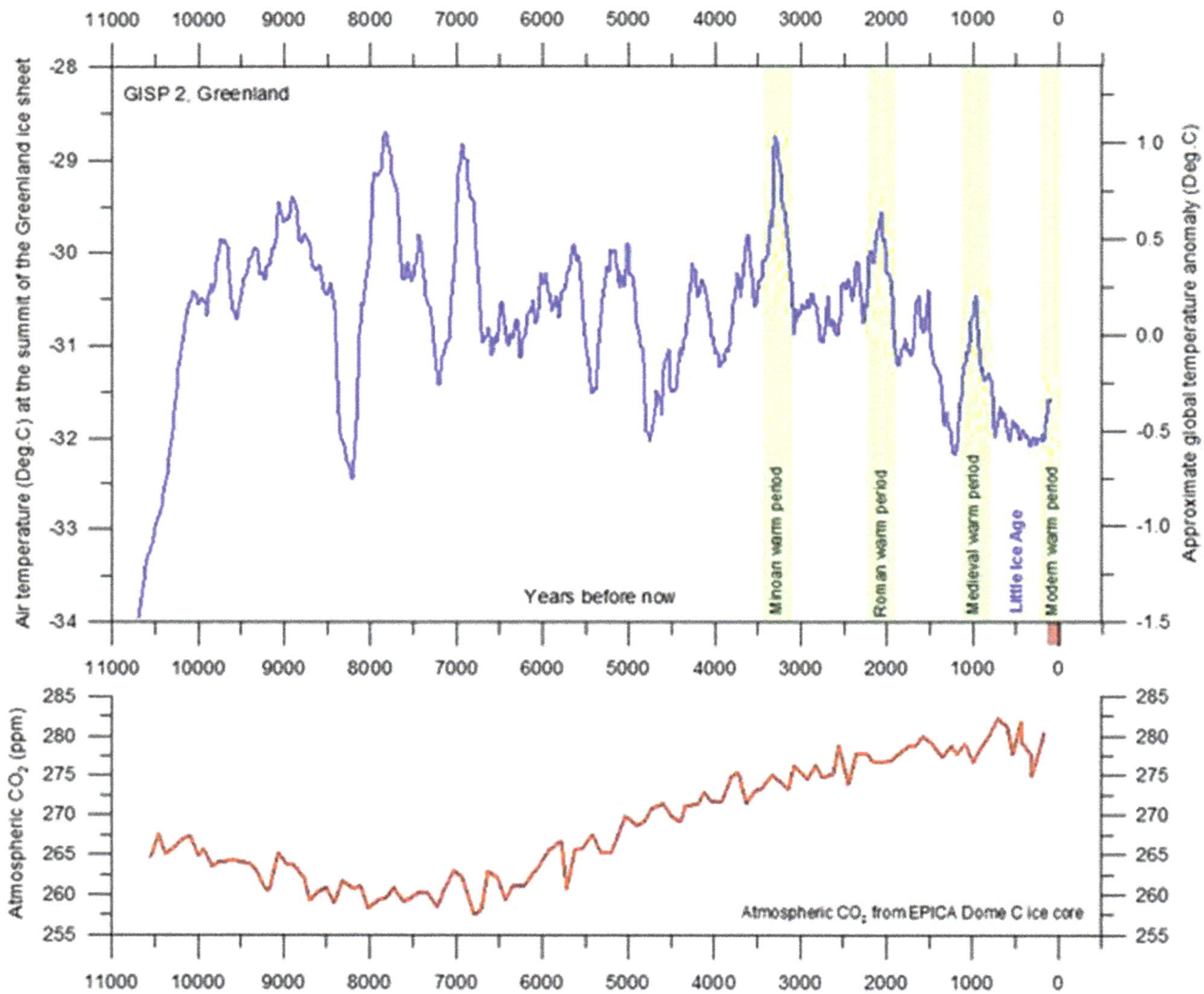

Figure 24. Global temperatures compared to CO_2 in the atmosphere during the Holocene warm period over the last 11,000 years. Source: Climate4you, Dr. Ole Humlum.

Temperatures and CO_2 concentrations over the last 11,000 years measured to approximately 1900, as determined from ice cores taken from the Greenland Dome C ice cores, show several things. First, there have been several much warmer periods in the Earth's recent past of the Holocene warm period than we have seen over the last period of time when mankind's industrial activity was producing CO_2. These include from the top chart, the Minoan warm period, the Roman warm period, and the Medieval warm period. Second, over the last 7,000 years, the atmospheric CO_2 concentration has been steadily climbing while the planet's temperatures have been dropping significantly. Both of these facts demonstrate that mankind's CO_2 production has little to do with the natural cycles of the Sun, and as recent research has shown, the Earth may be relatively insensitive to how much CO_2 is in the air. These facts and the research disclosed in *Dark Winter* make a strong case for believing that the Sun drives climate changes on Earth. It further reinforces that mankind's industrial CO_2 may have little to do with major climate variations.

The Analysis of California Earthquakes vs. Solar Hibernations

Similar to the solar hibernation chart for the NMSZ, the chart below gives us a clear picture of when the past periods of reduced solar activity were using sunspot counts. Hopefully by understanding when the next hibernation arrives, we will also know when the next devastating quakes will come if there is a strong correlation between sunspots and earthquake events of the most destructive magnitudes.

In the following chart, we have overlain the sunspot curve with 22 major California earthquakes since 1811, the date of the first reliably recorded quake. Here is the sunspot chart:

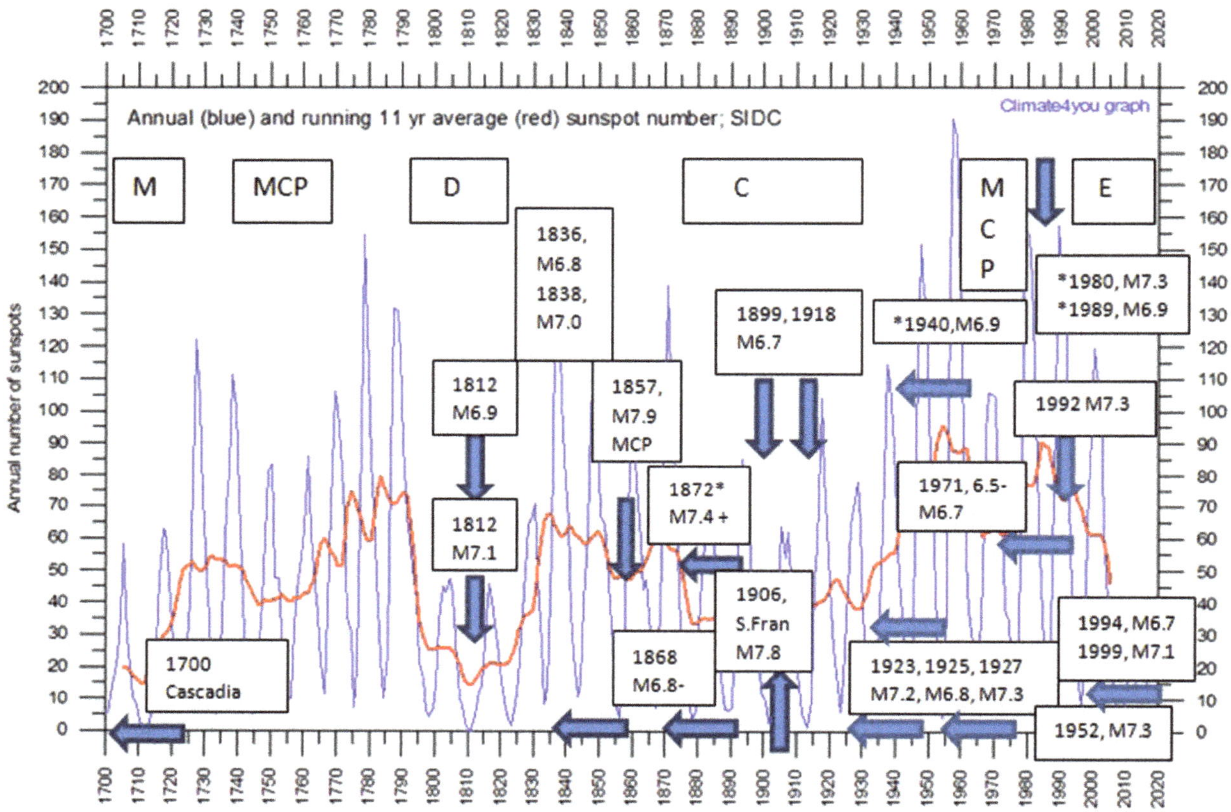

Figure 26. History of California earthquakes (M6.7+) compared to solar activity from 1700 to 2010. Source: J Casey Climate4you.com, Dr. Ole Humlum. This chart is packed full of relevant information. The red line indicates the smoothed average of sunspots. The thin blue line indicates sunspot counts for each of the normal 11-year solar cycles since 1700. Dates and magnitudes for each M6.7+ quake are indicated in text boxes. Positions on the sunspot time line for each of these quakes are shown by blue arrows. Note: the 1700 Cascadia mega-quake of M9.0+ is included for reference only. It occurred at the lowest point in solar activity in the last 317 years, the bottom of the Maunder minimum, the coldest years of the Little Ice Age. Note the various solar minimums in this chart which will be referred to throughout this book: The *M* for the Maunder minimum, the *D* for the Dalton minimum, the *C* for the centennial minimum, *MCP* for each of the multicycle pauses, and the *E* for the Eddy minimum. The USGS does not post reliable records for California quakes prior to 1812.

This chart is striking in its relevance to the asserted correlation between solar activity and earthquakes. This adds much weight to both the accuracy of the RC theory and the prediction of future major earthquakes. Here is what we have found when we examine past catastrophic California earthquakes of magnitude M6.7 or larger:

1. Of the 22 major earthquakes (M6.7+) tracked since 1812, 18 of 22 or 82% took place during either during a major reduction in solar activity (low points in red trend line) or between 11-year solar cycles when the sunspot counts were lowest and often near zero.

2. This 18 of 22 total includes eight of 22 or 36% of the earthquakes that took place between 11-year solar cycles when sunspot counts approached zero.

3. A total of 82% of all major California earthquakes took place during a sunspot low period.

4. Only four of 22, or 18%, did not take place during low solar activity including in 1827, 1940, 1980, and 1989.

The use of C14 data beyond 1950 is typically not desired since the cold war days saw the United States and the former Soviet Union creating an artificial spike in C14 in the atmosphere through their joint detonation of above-ground nuclear weapons tests.

The issue is moot in this text since we have charts depicting the next solar hibernation via sunspot proxies for the Sun's activity and therefore can identify the highest risk periods for future CGE from those charts without attempting to extend the C14 charts to the 2030s.

The Multicycle Pause

Of special interest to those of us who study solar activity is that the Ft. Tejon quake that occurred at the solar low point of a multicycle pause (MCP). The 11-year solar cycle during which the M7.9 quake struck was a particularly important solar event. There are three MCPs noted on the chart above. It was as if the Sun had just expended a lot of energy in a buildup phase of stronger and stronger 11-year cycles and then decided to cut back on its energy output, sunspot count, etc., and effectively catch its breath. This phenomenon is common during either a 206-year cycle or a 90–100-year or centennial cycle. It represents the central period of that particular cluster of multiple 11-year cycles. The MCPs on the chart occurred around

1750, 1855, and 1970. These pauses, easily seen off the chart above, happened approximately every 100 years and again speaks to the reliability and repeatability of solar activity cycles. Cold climate effects were felt for almost two decades either side of the 1970s pause, an almost 40-year global cooling event during which the global industries were setting new records yearly for increased greenhouse gas (CO_2) emissions, yet temperatures declined. Again the greenhouse gas theory simply does not work. Climate variation and global cooling are easily explained by solar cycles in the vast majority of climate changes we have seen over the past 2,000 years. The most important conclusion we can draw from the past earthquake history for the West Coast of the USA is this:

While the U.S. West Coast has a history of numerous large earthquakes, we know with a high degree of certainty (82%) that it does have major earthquakes during solar hibernations, solar minimums, or during the low sunspot period between 11-year solar cycles.

In fact, of the four largest earthquakes along the West Coast over the past 400 years, all four (100%) occurred at the very bottom of a solar minimum or pronounced drop in sunspot trend count. These were the Cascadia M9.0 quake and tsunami of 1700, the M7.9 1857 Fort Tejon quake, M7.8 1892 Imperial Valley quake, and the M7.8 Great San Francisco quake of 1906.

It is because of the link with solar minimums that I wrote letters to the U.S. government and the appropriate state governors that the highest risk window for these catastrophes to occur is between 2017 and 2038, the next such minimum. Since the calculation for the bottom of the next 206-year cycle is the year 2031, that means we have an even smaller time frame for California catastrophic earthquakes than we saw with the NMSZ. We should now be planning for catastrophic West Coast earthquakes between 2017 and 2031. This is a very narrow fourteen years—a blink of the eye in solar cycle time, much less on geological time scales.

Planetary/Crustal Energy Flow and Its Relationship on the Schedule for Catastrophic Earthquakes for California

The figure below provides a visual understanding of the progressively closer earthquakes that are approaching the latitude of California based on the energy flow analysis by Dr. Choi and Dr. Tsunoda, discussed above pertaining to the CSZ.

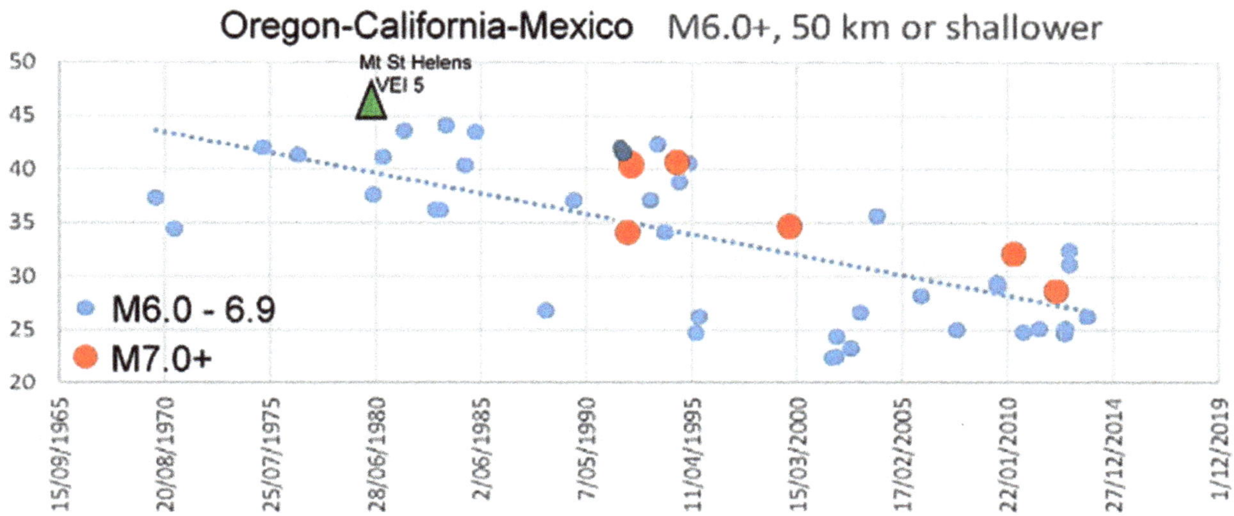

Figure 26. Chart of time vs. latitude for West Coast (California) earthquakes. This chart identifies an ever-increasing threat of major shallow earthquakes since the late 1960s. Source: D. Choi and F. Tsunoda.

This important chart indicates that by 2019, deep energy flows from within the Earth will have reached Southern California. These expected earthquakes between M6.0 and M7.0, possibly M8.0, will have the potential for catastrophic damage within the state, especially Southern California and, by extension, anywhere along the San Andreas Fault. This chart also shows as we have already found from sunspot analysis that a swarm of major earthquakes are likely rather than a few isolated temblors.

Arriving at the same general, though disconcerting threat of catastrophic quakes for the U.S. West Coast reinforces the value and credibility of the science behind the relationship between the Sun and the Earth's largest earthquakes.

This high-risk period can be refined even further by identifying the low sunspot periods between 11-year solar cycles. The low sunspot period between solar cycle 24 and 25 is expected to be approximately 2019–2022. This is when the energy flows as predicted by the scientists above approaches 20–25 degrees north latitude. The sunspot low point between solar cycle 25 and 26 should be approximately 2031–2033. The start to end time frame for both of these cycles is expected to be 2019–2042.

There is a history of major earthquakes in California that is strongly correlated to significant reductions in solar energy output and time dependent energy flows within the planet. Accordingly, we conclude that:

1. There is a greater than 80% probability of more than one catastrophic earthquake (M6.7+) striking the state during the time frame 2017 to 2033.This is consistent with the most recent USGS estimates.

2. We further believe that given the long cycles of the Sun and their association with the most destructive quakes that there is a greater than 80% chance of an M 7.9 or greater earthquake to strike the state in the 2017–2033 time frame.

3. Given the normal deviations that exist around the mathematically calculated time frames for such earthquake predictions, the planning for the worst possible earthquake events for catastrophic California earthquakes should include the entire high risk period of 2017–2033.

****

In summary, we cannot understate the severity of the warnings in this chapter for residents of the states of California, Oregon, and Washington. The long intervals between catastrophic earthquakes are a major sign of things to come. The cycles of the Sun examined by Dr. Humlum and myself along with and the energy flow analysis by Dr. Choi and Dr. Tsunoda have reinforced our assessments to a high level of certainty that a new epoch of catastrophic earthquakes has begun.

The many decades of relative comfort and safety from catastrophic earthquakes, enjoyed by our citizens along the U.S. West Coast, are over.

Regrettably, commercial development of high-rise buildings within the state, notably in Los Angeles and San Francisco, seem to run counter to this threat. As this book was in final edits on September 23, 2016, a disturbing article regarding a leaning, sinking 58-story building (the Millennium Tower) which opened in 2009 in San Francisco was printed in the *Wall Street Journal*. This was preceded on September 6, 2016 by a photo of a 73-story building (the Wilshire Grand Tower) in Los Angeles nearing completion. When finished, it will be the tallest building west of the Mississippi River. Are these buildings rated to withstand the M6.8-to-M7.9 range of predicted earthquakes between 2017 and 2038? What about an M8.0?

The sooner that residents and local and state governments take these warnings seriously, the more lives can be saved, the death toll minimized, and post-quake recovery can be of optimum effectiveness and speed.

Chapter 5

Alaska: Big State, Big Earthquakes

We live in an epoch when the solid ground beneath our feet shakes daily.
—Barbara Ward

This chapter was begun with some trepidation, some intellectual shaking beneath the feet. All the comparative analyses done in this book between solar activity and earthquakes within various regions of the United States have produced consistent results thus far. They have pointed time and again to the purity and reliability of the both the RC theory. Still, this chapter about the great state of Alaska with its many geophysical nuances created some small apprehension for the first time about the outcome. The prospect that Alaska would be the undoing of our methodology was the source of the state of unease. The matter reached an apex of distress with our first plot of the first major Alaskan earthquake against the last 150 years of recorded sunspot trends.

Barbara Ward's summary of the uncertainty of life and its shaking-ground metaphor is not only spot on given the subject of this book but also fully reflects the history of Alaska and its long list of major earthquakes—far more than any other U.S. state. It also drives home the paired anticipation of both success and failure every researcher confronts with the crunching of the numbers from first new set of test data or analysis.

The angst comes from the fact that the geology underlying Alaska and its location on the planet is quite a bit different than that of the continental United States, including the West Coast, the NMSZ, or even the Cascadia Subduction Zone.

After all, the Aleutian megathrust fault or let's call it the Aleutian Subduction Zone (ASZ) is 90 degrees out of phase with regard to the Cascadia Subduction Zone and separated by more than the length of the entire state of California. That is, the CSZ runs essentially south to north along the West Coast. The ASZ runs east to west. Most importantly since it already lies within some of the coldest latitudes, cold climate effects on the Earth's crust, if they have a say in seismic activity, would seem to be minimal or without a clear trend. In addition, Alaska is a state with numerous moderate, large, and catastrophic earthquakes and volcanic eruptions. It is a veritable smorgasbord of geophysical events. Where we are all rightfully concerned with a possible M8.0 striking the LA basin with the last one being 160 years ago, Alaskans snicker at such a quake history. Alaska has had at least eleven catastrophic quakes of M7.9 or larger in just the last 117 years, and one of those was an M9.2! California for comparison has had just one—the M7.9 1857 Ft. Tejon quake.

The difference of course is one of relative damage and loss of life. An M8.0 today in Los Angeles could produce many times the number of deaths of all eleven quakes to have hit Alaska over those 117 years. Alaska, with only about 739,000 residents spread across its vast 587,000 square miles of landmass, simply doesn't compare with California, the most populous state in the United States, in terms of what a catastrophic quake could deliver in numbers of fatalities and injuries.

Figure 29. Map of Alaska. Source: State of Alaska.

Alaska is almost one-fifth the total size of the 48 states. Here is the comparison to the 48 contiguous states.

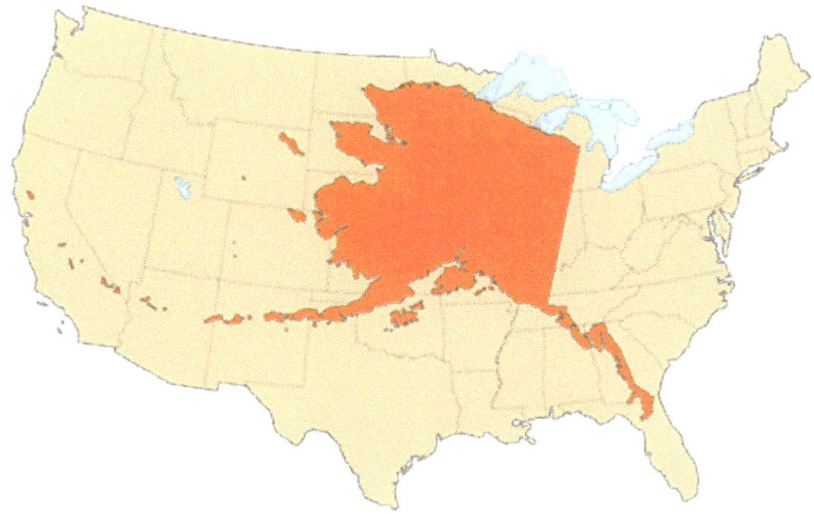

Figure 29. Comparison of the size of Alaska. At 586,412 square miles to that of the 48 contiguous U.S. states at 3,119,885 square miles. Source: Eric Gaba – Wikimedia Commons user: Sting.

This former Russian territory was sold to the United States during the Andrew Johnson administration and was championed by Secretary of the Treasury William Seward. Many decades later, the wisdom of Seward's purchase has been proven in many ways to have been a great investment in view of the natural and mineral resources the state has produced.

Alaska is a vast wonderland for nature lovers, rugged outdoorsmen and women, or those who want to get away from it all amongst towering mountains and numerous waterways, abundant forests, and ample wildlife. Its state capital, Juneau, is far south of the main body of the state much of which lies at the same northern hemisphere latitudes as Greenland, Iceland, Norway, Sweden, and Finland. It is the only U.S. state that lies partially within the Arctic Circle. Its winters are the stuff of legend. Alaskans are a tough breed by nature and because of nature.

It is also the U.S. state with a history of the most massive earthquakes of any U.S. state over the past century. Perhaps the best known is the M9.2 quake that struck on March 27, 1964.

As the map below shows, Alaska lies at the northern part of the Pacific Ocean, and its portion of the Aleutian Islands are little more than a chain of ancient islands that produce quite active volcanoes, large earthquakes, and tsunamis.

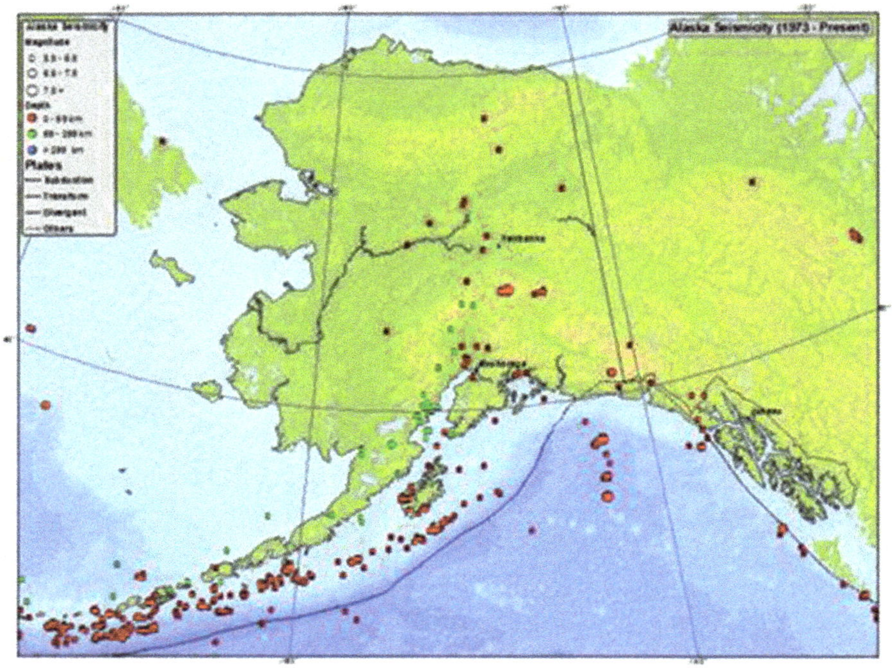

Figure 29. Locations of Alaskan earthquakes since 1973. Source: USGS. Looking at the state like a gigantic comma, the head is the large center part with the tail of the comma out to the west in the Pacific Ocean being the Alaskan Peninsula. Farther to the west are the Aleutian Islands which reach virtually to Russian Siberia. The long blue line along the island chain is the Alaskan Subduction Zone (ASZ).

The general high risk areas for earthquakes in Alaska are depicted below.

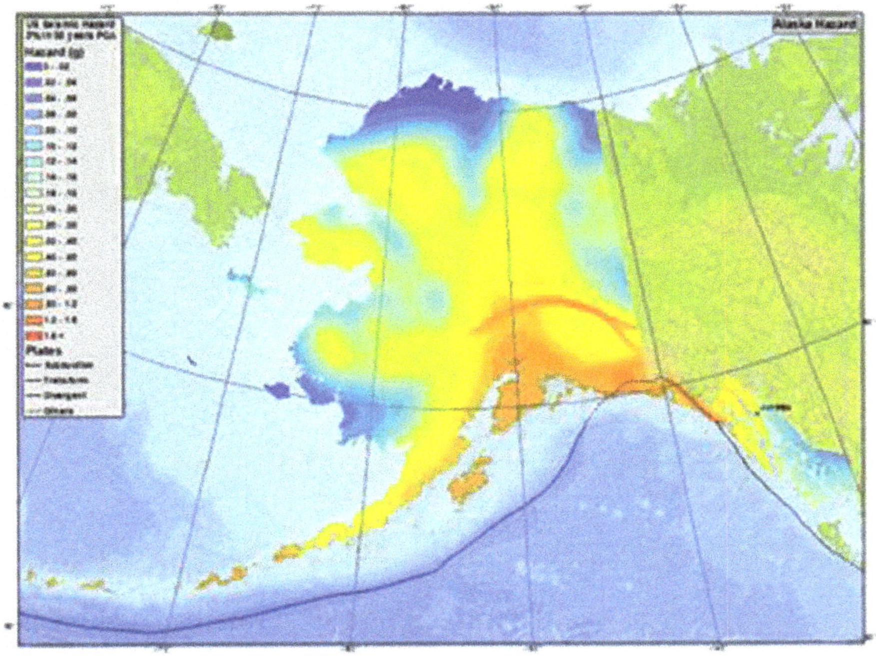

Figure 30. Alaska earthquakes hazards map. Yellow and orange indicate the areas of greatest earthquake threat. Source: USGS, 2007.

From this USGS chart, we can see that the Aleutian Islands, the southern coast, and the center of Alaska are high-risk zones. The dark orange line runs along the Denali Fault which lies just south of Fairbanks. The dark blue line is the Aleutian megathrust fault zone or the ASZ. The state's southeastern leg goes down along far western Canada and the far western extension of the Aleutians borders Russian territory. The Bering Strait off the northeastern coast of Alaska is within eyesight on a clear day from Russia's Siberian peninsula.

Alaska's vastness, challenging weather, its low population density combined with its natural ruggedness, earthquakes, and volcanoes, makes Alaska truly unique and almost beyond comparison with any other U.S. state. The top seven of ten most powerful U.S. quakes this century were in Alaska, not California. The list below from the Alaskan Earthquake Center at the University of Alaska–Fairbanks is just the major quakes that have struck Alaska since 1964:

- 2015 M6.4 Iliamna Earthquake
- 2015 M6.9 Fox Islands earthquake
- 2015 M6.8 Sutwik Island Earthquake
- 2014 M6.3 Skwentna Earthquake
- 2014 M5.2 Minto Earthquake
- 2014 M6.0 Palma Bay Earthquake
- 2014 M6.0 Seward Glacier Earthquake
- 2014 M7.9 Rat Islands Earthquake
- 2014 Noatak Earthquake Swarm
- 2013 M7.0 Andreanof Islands Earthquake
- 2013 M7.5 Queen Charlotte Fault Earthquake
- 2012 M5.8 Northern Cook Inlet Earthquake
- 2012 M6.3 Gulf of Alaska Earthquake
- 2012 M6.4 Andreanof Islands Earthquake
- 2012 M6.2 Fox Islands Earthquake
- 2011 M7.3 Fox Islands earthquake
- 2011 M5.2 Kantishna Earthquake
- 2010 M6.7 Fox Islands Earthquake
- 2010 M6.5 and M6.3 Bering Sea Earthquakes
- 2009 M6.5 and M6.4 Fox Islands Earthquakes
- 2009 M5.4 Skwentna Earthquake
- 2008 Kasatochi Earthquake Swarm
- 2008 M6.6 Andreanof Islands Earthquake
- 2008 M6.6 and M6.4 Andreanof Islands Earthquakes
- 2007 M6.4 Fox Islands Earthquake
- 2007 M7.1 Andreanof Islands Earthquake
- 2007 M6.3 Alaska Peninsula Earthquake

- 2007 M6.4 Andreanof Islands Earthquake
- 2007 M5.7 Southwest Yukon Territory Earthquake
- 2007 M6.7 Andreanof Islands Earthquake
- 2007 Northeast Brooks Range Earthquake Swarm
- 2006 M6.6 Andreanof Islands Earthquake
- 2006 M6.4 Rat Islands Earthquake
- 2006 M6.4 Fox Islands Earthquake
- 2006 M5.4 Yukon Flats Earthquake
- 2005 M6.2 Unimak Island Earthquake
- 2005 M6.8 Rat Islands Earthquake
- 2003 M7.7 Rat Islands Earthquake
- 2002 M7.9 Denali Fault Earthquake
- 1999-2001 Kodiak Island Earthquakes
- 1996 M7.9 Adak earthquake
- 1965 M8.7 Rat Islands Earthquake
- 1964 M9.2 Great Alaskan Earthquake

This list of major quakes is unseen in the records of any other U.S. state. Because of the remoteness of Alaska, the records of major quakes there don't go back far. If this had been the list of quakes in California since 1964, it's likely many of the state's citizens would have moved out long ago.

Figure 31. A map of the great 1964 Alaskan M 9.2 earthquake and the epicenter (red star). Source: USGS.

Figure 32. Image of Fourth Avenue in Anchorage, Alaska, after the M9.2 great quake of 1964. Source: USGS

Let's find out then whether Alaska has the same climate correlation as do other major high-risk earthquake regions of the USA.

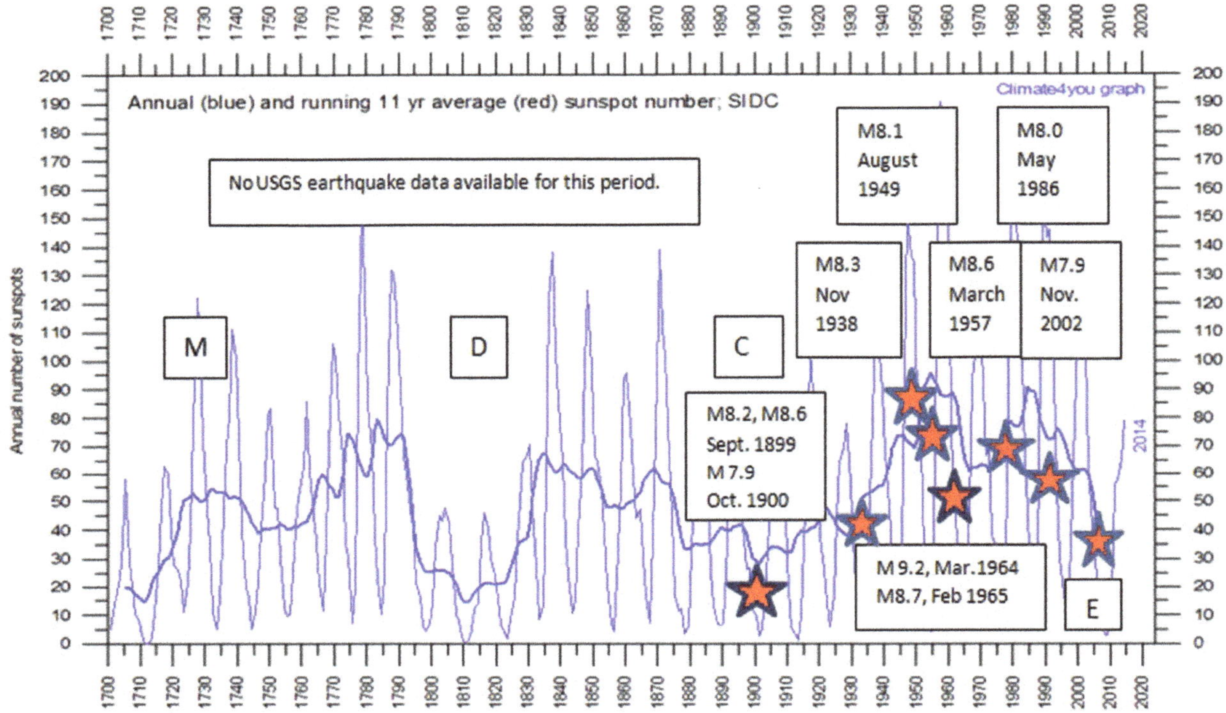

Figure 33. Major Alaskan earthquakes greater than M7.9 compared to sunspot records. The text blocks signify the *M* or Maunder minimum, the *D* or Dalton minimum, the *C* or centennial (Gleissberg) minimum, and the initial phase of the *E* or Eddy minimum. The light blue lines or spikes represent individual 11-year sunspot cycles. The dark blue line represents the sunspot average trend line—the key line to follow for sunspot trends. The final star to the right in the chart represents the Rat Island quake of June 23, 2014 which occurred during the current solar cycle 24, a time of rapidly declining solar activity noted by the dark blue line. Sources: J. Casey and Dr. Ole Humlum, (Climate4you.com), quake data from the USGS.

The results of plotting major earthquakes (M7.9+) since 1899 show that the 11 catastrophic earthquakes to hit Alaska and the Aleutians reveal strong correlations to solar activity:

1. Seven of eleven (64%), including the largest (M9.2, 1964), occurred at the bottom of a sunspot low point or recognized major solar minimum or between solar cycles when the sunspot count was near zero. They include the M8.2 and M8.6 quakes of 1899 and

the M7.9 1900 quake, the M9.2 1964 and M8.7 1965 quakes, the M7.9 2014 quake and the 1986 M8.0 quake. With the exception of the 1986 quake, they all happened during either the 1900 centennial minimum or within the multicycle pause (MCP) centered around 1970. The 2014 M7.9 Rat Islands quake which happened during the initial phase of the 206-year solar hibernation and the Eddy minimum is right on schedule with the Choi-Maslov relationship's brief increase in solar activity seen in 2014—the peak of the current solar cycle 24.

2. Another 2 of the eleven (18%) occurred during a long-term trend of declining sunspots. They include the M8.6 1957 and the M7.9 2002 quakes.

3. Only two of the eleven (18%) took place during an increasing trend or peak of sunspots and do not support the premise of major quakes during declining solar activity. They include the M8.3, 1938 quake, and the M8.1, 1949 quake.

4. Nine of the eleven earthquakes catastrophic earthquakes (M7.9+) or 82% to hit Alaska since 1899 happened during a declining sunspot trend or pronounced solar minimums.

5. Therefore, the next solar minimum, the Eddy minimum, centered around the year 2031 is the most likely period for an M7.9 or greater earthquake to strike Alaska.

The important time frame analysis based on past major quakes relative to associated solar low points was conducted. It indicates:

The time frame within the Eddy minimum which represents the highest risk to Alaskans is the period from 2022 to 2040 based on the nine of the eleven catastrophic past earthquakes that occurred during a solar minimum period.

In addition, the Aleutians and Alaska are affected by a crustal energy flow defined by Dr. Choi and Dr. Tsunoda and are explained in appendix 2. As shown in the chart below extracted from that paper, the earthquake and volcanic activity roughly follow a migration of events that can be used to identify the next highest risk period by latitude. The important conclusion from this chart of crustal energy flow is that by and beyond 2015, from the Aleutians heading toward eastern Alaska, the threat of major earthquakes and volcanic eruptions grows.

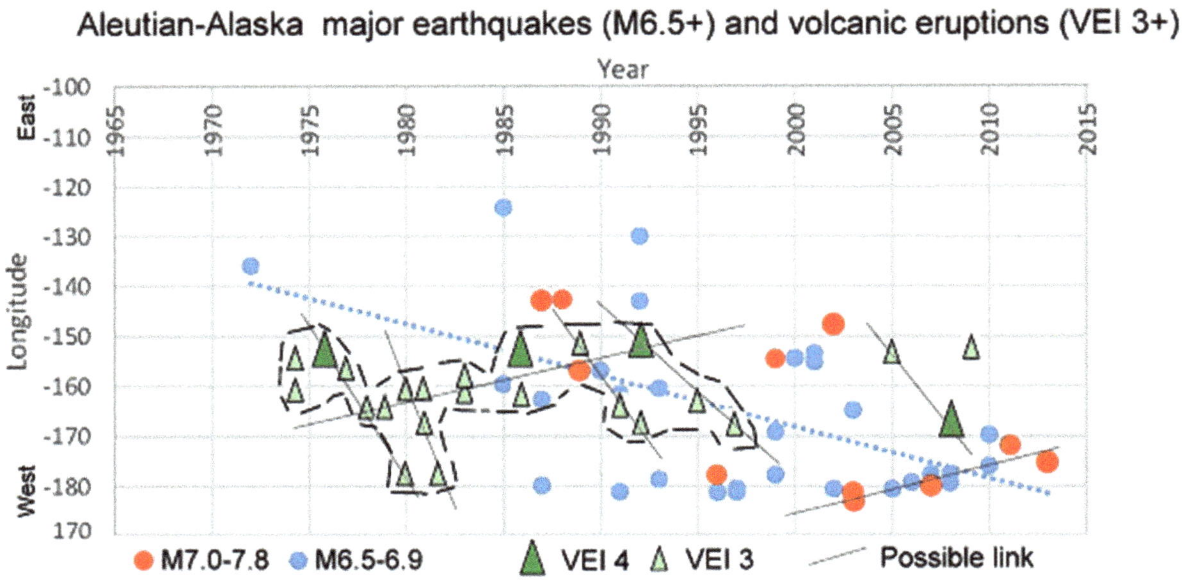

Figure 34. Longitude vs. time plot of major volcanic eruptions and earthquakes. Overall trend shows the westward movement of volcanic and seismic energies. But also seen is the eastward propagation in both volcanic eruptions and earthquakes. Source: Dr. Choi, F. Tsunoda.

The evidence speaks loudly that despite its special geology and latitude, the Alaskan state, including the Aleutians, do in fact follow the RC theory to a >80% correlation when it comes to the very largest earthquakes that have struck there since 1899.

Trepidation ended.

Chapter 6

South Carolina, Puerto Rico, and Hawaii

Practice yourself in little things, and thence proceed to greater.

—Epictetus

One might ask why include South Carolina, Puerto Rico, and Hawaii in any earthquake analysis? They only have little earthquakes and then not often. Is this so? Or is there some greater understanding we can derive from a look at these often less-discussed earthquake risk areas? We have found the results are surprising.

Further, the more areas we examine for past trends and patterns of earthquake behavior, the more that will be revealed which we are not already cognizant of on the broader picture of earthquake characteristics. The study of seismology or other fields suggests this not unexpected. That is the true quest of science, in an effort to come up with the best theory and best results possible. It is a constant process of refinement.

The Earthquake Threat for South Carolina

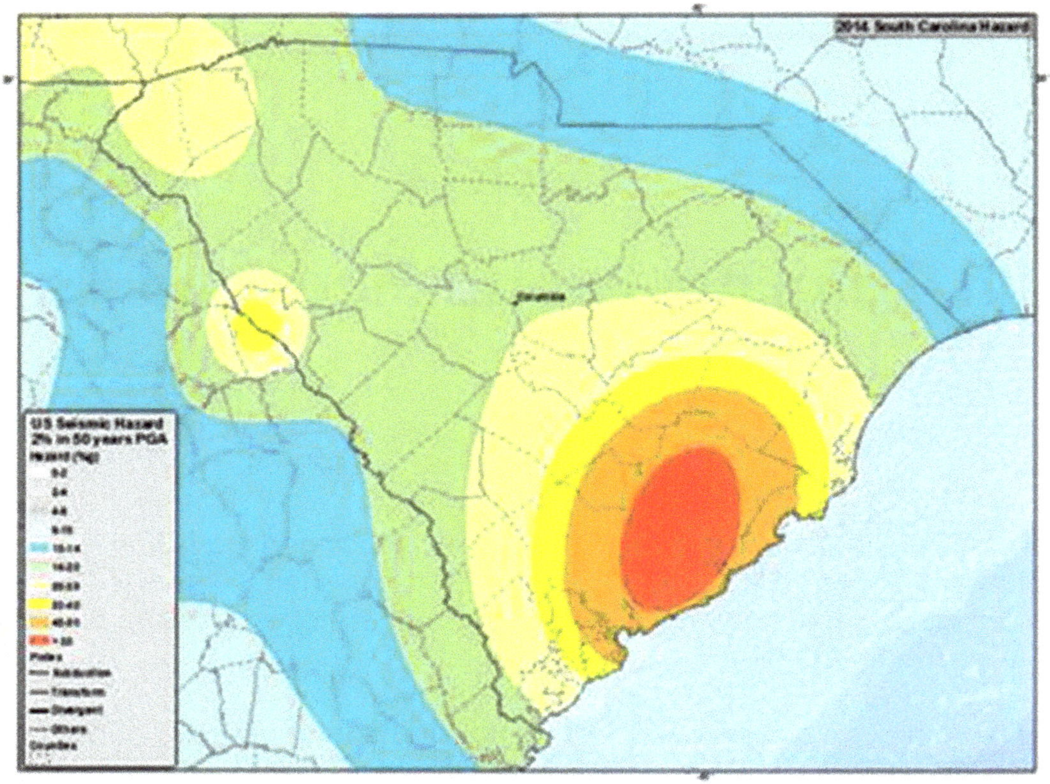

Figure 35. Seismic hazard map of South Carolina. The bright red section of the graphic shows where the greatest earthquake threat lies within the state of South Carolina. It is centered on the city of Charleston. Source: USGS.

On August 31, 1886, the state was rocked with the largest earthquake ever recorded along the Atlantic coast of the United States. The M7.3 temblor killed over 60 while destroying most of downtown Charleston's brick structures. The quake was felt from Cuba to Boston.

Figure 36. Damage to a building in Charleston, South Carolina from the 1886 M 7.3 quake. Source: USGS.

A 2001 study for the State of South Carolina titled "Comprehensive Seismic Risk and Vulnerability Study for the State of South Carolina" prepared for the South Carolina Emergency Preparedness Division had the following damage assessment should another M7.3 strike like the 1886 event. The study also looked at lesser-postulated quakes under different scenarios involving M6.3, M 5.3, and M 5.0 quakes. Here are damage assessments in the event of a repeat of the 1886 event:

> Results from the M 7.3 scenario include: Economic losses due to building damage alone are estimated to be over $14 billion (2000 dollars) with ground failure effects included, compared to the $2 billion for the M 6.3 event. Losses to lifelines would result in more than $1 billion for the M 7.3 event. About $10.9 billion or about 77 percent of the total economic losses will occur in the Tri-County region (Charleston, Berkeley, and Dorchester Counties). The building damage alone will cause over $4.2 billion in losses due to business interruption in the State. These losses correspond to rental income losses, lost business income, wage losses, and expenses associated with relocation. Secondary business interruption losses related to lost revenues to suppliers and wholesalers are not included. A daytime event will cause the highest number of casualties. Of the estimated 45,000 casualties, close to 9,000 or about 20

percent will be major injuries (injuries requiring hospitalization) and fatalities (about 900). Most of these casualties will occur in Charleston, Dorchester, and Berkeley counties. Nearly 70,000 households, or about 200,000 people are expected to be displaced, with an estimated 60,000 people requiring short-term shelter. Fire following a M 7.3 earthquake in the Charleston area will be concentrated primarily in the Tri-county region. The scenario earthquake is expected to cause over 250 fires. The lack of operational firefighting equipment and a supply of water for fighting fires after a large earthquake may become a major concern in effectively fighting these fires. Due to insufficient seismic building code standards and the vintage of the building stock, the majority of the structures in the State, in particular schools and fire stations are vulnerable to damage. Indeed, it is estimated that over 220 schools (not considering the extensive damage to the relocatable school buildings) and over 100 fire stations will experience significant damage. This may lead to some potential issues with respect to providing reliable shelters for immediate use in emergency response and sheltering and with respect to responding effectively to the 250 fires, expected from this scenario. Schools are expected to suffer significant damage in the case of the M 6.3 scenario, as well. Furthermore, there could be some safety issues related to school children, teachers, and other persons in school buildings. The catastrophic failure or partial collapse of one or more school buildings during school periods could greatly increase the casualty estimates. Restoration of the schools for the emergency sheltering of the homeless and other contingency service will be demanding. Over 36 million tons of debris will be generated, including an estimated 10 million tons of Category II debris, which includes concrete and steel – materials that require special treatment in "deconstruction" and disposal. Debris disposal, therefore, may pose a major challenge in the recovery phase. This total does not include biomass. Hospitals will likely suffer significant building damage that could result in more than 30 hospitals out of the 108 (about 30%) being nonfunctional. Over half of these affected hospitals may experience extensive damage. The M 6.3 event will result in about 10 hospitals suffering considerable damage. Since most of this damage will be concentrated in the Tri-county area, the region may be faced with the serious issue of how to provide the needed care to existing patients and potential thousands of earthquake victims from the affected communities. Close to 800 bridges are expected to suffer enough damage to make them inaccessible, thus, hampering even further the recovery efforts. In addition, certain communities in the greater Charleston area are that are only accessible by bridge routes may

> be cut off. A good portion of the Charleston area is susceptible to liquefaction. However, ground failure effects contribute only about 5% or less to losses. Of all the utility systems, electric power is arguably the most critical, as many other lifelines depend on it. It is expected that about 63 electric power facilities, (51 substations out of the total of 380 and 12 power plants out of the total of 53) will suffer at least moderate damage and nearly 300,000 households will be without power, right after the earthquake. In potable water pipes greater than 12 inches, over 1100 repairs will be needed, or about a repair for every two kilometers of these pipes. Over half of these are expected to be breaks. Widespread water failure may drain water within minutes or hours from the distribution system, thus preventing adequate water supply for fire suppression. In addition, about 80% of the urban households in the affected area will be deprived of water. It will take weeks, if not months, to restore the serviceability of the water systems. Therefore, significant external augmentation would be required to provide and sustain such a high repair level. In the event of a M 6.3 earthquake in Charleston, approximately 136,000 buildings will sustain slight to moderate damage and 25,000 will be extensively damaged. Total building loss including capital stock and income losses will exceed $2 billion. Approximately 30 to 60 people will be killed and from 2,000 to more than 3,000 people will suffer minor to major injuries. In the M 5.3 Charleston scenario earthquake, the losses and casualties decrease significantly. Injuries will number less than 100 with no estimated deaths. Total loss to buildings will be about $230 million. If a small earthquake of M 5.0 were to occur in Columbia, approximately 400 buildings would sustain slight or moderate damage with a total loss of $310 million. Less than 10 people will be injured and only with minor injuries. In summary, a repeat of the M 7.3 Charleston earthquake in South Carolina, at least in the early aftermath, may cause the State to be overwhelmed by widespread damage as well as the disruption of lifelines. The impact from this event demonstrates the scope of the problem and reinforces the need to implement structural and non-structural mitigation measures as a central feature in long-term initiatives to reduce seismic risk. Affected communities will be coping with the trauma and demands of immediate response and early recovery.

The damage assessment from this relatively old study (16 years ago) is much that like from all other earthquake damage projections we have done across the United States. We see that the rapid population growth and land development of the intervening period has added greatly to the potential loss of life, suffering, and property losses we can expect from a repeat of the any earthquake trend that is 100 years or more older, including the 1886 quake.

The study above does much. But one thing it does not do is focus on one area of concern that was mentioned in our look at the NMSZ—the safety of nuclear reactor power stations. It discusses the electrical power matter for sure, but unfortunately, the study does not mention nuclear reactor power facilities or even use the word *nuclear* or *nuclear power* in the study conclusions or recommendations. It does not evaluate the safety of these facilities or whether they will be able to withstand another moderate M 5.5 like the 1913 quake in the foothills of the Appalachians and says not a word on whether any nuclear facilities in the state are safe from the direct and indirect effects of another M7.3 quake. Nuclear facility safety since its inception has been with the Nuclear Regulatory Commission (NRC) and the U.S. Department of Energy (DOE). States often just have to wash their hands and say it's a federal concern.

The practical truth is that it is everyone's concern, and the federal government's responsibility to make sure such facilities are safely operated and, in our case, are not subject to the whims of Mother Nature. It would be reassuring if the U.S. government has gone to every high-risk earthquake zone of the country and recertified each nuclear power plant and other facilities housing and processing radioactive material to verify they pose no risk to the public from a catastrophic earthquake perspective. Is there anyone in the administration willing to step forward and state that all such facilities can withstand the worst-case scenario?

We have no confidence whatsoever that our nuclear industry, power stations, etc., are rated for the maximum earthquake(s) anticipated for their particular state or region. Like all other aspects of the current U.S. government and in particular this administration, we believe that this is another functional area of the U.S. government that will someday, perhaps soon, show us that it is also poorly, possibly tragically mismanaged. We should expect that in South Carolina, the NMSZ, or other earthquake-prone regions, the NRC and the DOE have failed to do what is needed to protect the people from the next twenty years of catastrophic geophysical events.

Until independent observers have checked these nuclear facilities, we should assume that all nuclear facilities in the United States are not able to handle the direct and indirect effects of catastrophic earthquakes we predict in this book and definitely not in time to be safe before the window of highest risk opens. That by the way is 2017!

Another valuable tool for understanding the seismic threat to South Carolina is to look only at the area of Charleston, the site of the only catastrophic or even moderate quake in the state's history. Though the state has had many quakes during its recorded history, including the M5.5 in 1913, our look at how to evaluate the South Carolina earthquake threat comes down to this. The total of the state's quakes since 1698 were all minor ones in which little damage or deaths were reported. Thus, we can conclude the threat to the state must be focused on the immediate Charleston area where the only catastrophic quake in the last 500 years has happened.

The data that supports this conclusion comes in part from research published in two reports:

1. Earthquakes in South Carolina and vicinity 1698–2009: U.S. Geological Survey Open-File Report 2010–1059, by Dart, R. L., Talwani, Pradeep, and Stevenson, Donald, 2010.

2. A PDF report on the prehistory of earthquakes in South Carolina by Talwani and Schaeffer. This study shows earthquake cycles in the state for CGE's begin at over 500 years. The next 1886-style quake may be in 2017 but more likely will be 150 years in the future according to the study authors.

Based on the history of quakes reported in the studies, we can plot the occurrence of a selected group of the five largest related to the primary threat area around Charleston to see whether there is, as we expect to find, a strong correlation to solar activity.

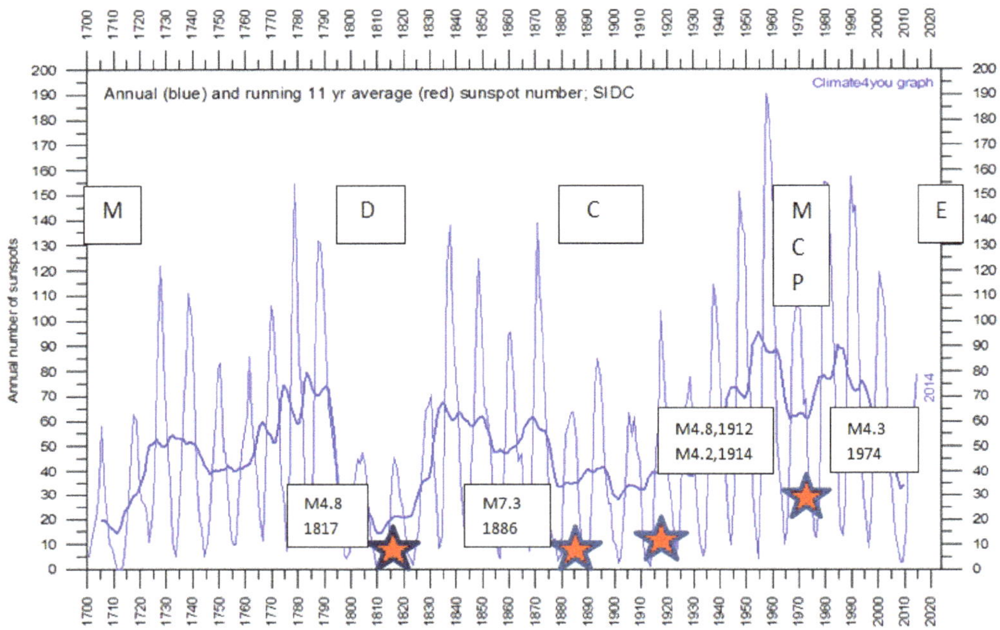

Figure 37. Charleston, South Carolina earthquakes compared to sunspot records. Note the solar activity low points for the *M* Maunder minimum, *D* Dalton minimum, *C* the Centennial minimum, *MCP* for the multicycle pause of the early 1970s, and *E* for the new Eddy minimum Source: J. Casey, Dr. Ole Humlum (Climate4you.com), Quake data from USGS—Dart, R.L., Talwani, Pradeep, and Stevenson, Donald, 2010.

The history of the five largest earthquakes in Charleston since 1700 are stunning in their relationship to solar activity. Here is what we can say about the possibility of the past and future for Charleston:

1. All of the top five past quakes (100%) greater than M4.2 struck during a major solar decline or hibernation of the Sun.

2. These past quakes took place during either a dramatic reduction in sunspots (MCP-1974) or during a centennial minimum, or during the Dalton minimum (a solar hibernation).

3. If the sequence of these past five earthquakes holds, this means that there is at least an 80% chance of the next quake (M4.2 to M7.3) occurring during the next steep drop in sunspots or during the Eddy minimum, the next solar hibernation.

Note that the M5.5 quake of 1913 does not factor into the Charleston-only quake assessment above, yet it falls inside the centennial sunspot low point and would be right alongside the other two centennial minimum Charleston quakes if it were included.

Important conclusions may now be derived for Charleston as we have for other earthquake regions in terms of the time frame of when the next significant quake will strike Charleston. We can determine the past range of each of the past five quakes from their associated sunspot low point (by year) and calculate the next range for Charleston quakes based on the next sunspot low point predicted to be 2031. Here are the results:

Year	Associated Low-Sunspot Year	Difference in Years	Comparative Year from 2031
1817	1810	+7	2038
1886	1900	-14	2017
1912	1900	+12	2043
1814	1900	+14	2045
1974	1974	0	2013

The table above combined with the plot of sunspots and earthquakes means that between 2017 and 2045 we have an 83% + chance of the next significant if not catastrophic earthquake in Charleston.

We obtain this 83% number by assuming the next major quake does not occur within the predicted time frame. That will leave us with a total 5 out of 6 periods that confirm the solar activity correlation, for a minimum quake strike rate of since 1817 of 83%. However, from 1817 to 2017, for the past 200 years, the quake rate compared to solar minimums has been 5 out of 5, or 100%. We expect nature to continue to be quite reliable in the future when solar activity and global cooling cycles are used as the primary tools for such predictions.

While there is little to be concerned about at the M 4.3 level of earthquakes, it cannot be determined that the expected next quake during the Eddy minimum will be equal to or greater than the 1886 M7.3 magnitude and thus highly dangerous to the citizens of Charleston.

Though we are dealing with a small data set, this region of the United States does not afford us much given the weak to moderate quakes found there which pose no threat to the populace. However, we should not dismiss the singular data point in that the most powerful quake—the M7.3 of 1886—took place a full 14 years prior to the sunspot trend low point. The rest of the four much smaller quakes took place at or after the sunspot low point.

Here's what that implies:

1. The range of years predicted for the next Charleston quake is 2017 to 2045—a difference of 28 years.

2. The previous catastrophic M7.3 quake took place 14 years prior to the solar low point of 1900.

3. Back-dating 14 years from the predicted Eddy minimum low point year of 2031 gives us 2017. Therefore, if history repeats, it would not be a surprise to see the most destructive quake for Charleston for the next 28 years to be a significant quake hitting the city as early as 2017!

South Carolinians, especially those living in Charleston, and the federal government should plan accordingly.

Earthquake Risk to Puerto Rico

With no specific data to back up this off-the-cuff hypothesis but in view of the results of countless man-on-the-street interviews from Jay Leno when he starred on *The Tonight Show* and Fox News (*Watter's World*), it is probable that the majority of Americans could not find Puerto Rico (PR) on a map of the world, Hispanics not included.

Except for the occasional story about abysmal state of the PR economy, its debt, and crime crisis, we would likely never see anything in our newspapers or on TV about it. Certainly the same can be said, even more so, with regard to its earthquake history. Yet the earthquake history of this small, warm, and humid country is of genuine importance in what it can tell us about its geology and whether major earthquakes can be expected to strike there any time in the immediate future.

Figure 38. Map of Puerto Rico. Source: Puerto Rico (October 28, 2016) in Wikipedia, the Free Encyclopedia. Retrieved 20:39, October 28, 2016 from https://en.wikipedia.org/w/index.php?title=Puerto_Rico&oldid=746671412

Here is what the USGS has to say about PR's past earthquake record and that of the nearby U.S. Virgin Islands:

> Four strong earthquakes have affected Puerto Rico since the beginning of its colonization. The most recent of these occurred on October 11, 1918. The epicenter was located northwest of Aguadilla in the Mona Canyon (between Puerto Rico and the Dominican Republic). This earthquake had an approximate magnitude of 7.5 on the Richter scale and was accompanied by a tsunami ("tidal" wave) which got up to 6 meters (19.5 feet) high. Damage was concentrated in the western area of the Island because this was the closest zone to the earthquake. The earthquake killed about 116 people and caused more than 4 million dollars of damage. Numerous houses, factories, public buildings, chimneys, bridges and other structures suffered severe damage.
>
> On November 18, 1867, 20 days after the Island was devastated by Hurricane San Narciso, a strong earthquake occurred with an approximate magnitude of 7.5 on the

Richter Scale. The epicenter was located in the Anegada Passage, between Puerto Rico and St. Croix, Virgin Islands. The earthquake produced a tsunami that ran inland almost 150 meters (490 feet) in the low parts of the coast of Yabucoa. This quake caused damage in numerous buildings on the Island, especially in the eastern zone.

Possibly the strongest earthquake that has affected Puerto Rico since the beginning of colonization occurred on May 2, 1787. This was felt strongly throughout the Island and may have been as large as magnitude 8.0 on the Richter Scale. Its epicenter was possibly to the north, in the Puerto Rico Trench. The quake was felt very strongly all across the Island. It demolished the Arecibo church along with the El Rosario and La Concepcion monasteries and damaged the churches at Bayamon, Toa Baja and Mayaguez. It also caused considerable damage to the castles of San Felipe del Morro and San Cristobal, breaking cisterns, walls and guard houses.

The other strong earthquake, whose magnitude has not been determined, occurred in 1670, significantly affecting the area of San German District." (Translated from "Terremoto," written by Jose Molinelli Freytes, Puerto Rico Civil Defense, under the auspices of the Federal Emergency Management Agency [FEMA])

Figure 39. Photo of 1918 earthquake damage in the Puerto Rico city of Mayaquez. Source: Wikimedia Commons, University of Puerto Rico (photo, Cifuentes).

In what may not have been a surprise, another strong earthquake struck Puerto Rico off shore on January 13, 2014. As mentioned before, this quake struck at the peak of the 11-year solar cycle 24 exactly when the Choi-Maslov relationship suggested global quakes might increase briefly. Fortunately, no deaths were reported.

Here is what PR's earthquake history looks like plotted once more against the activity of the Sun over the past four hundred years or so.

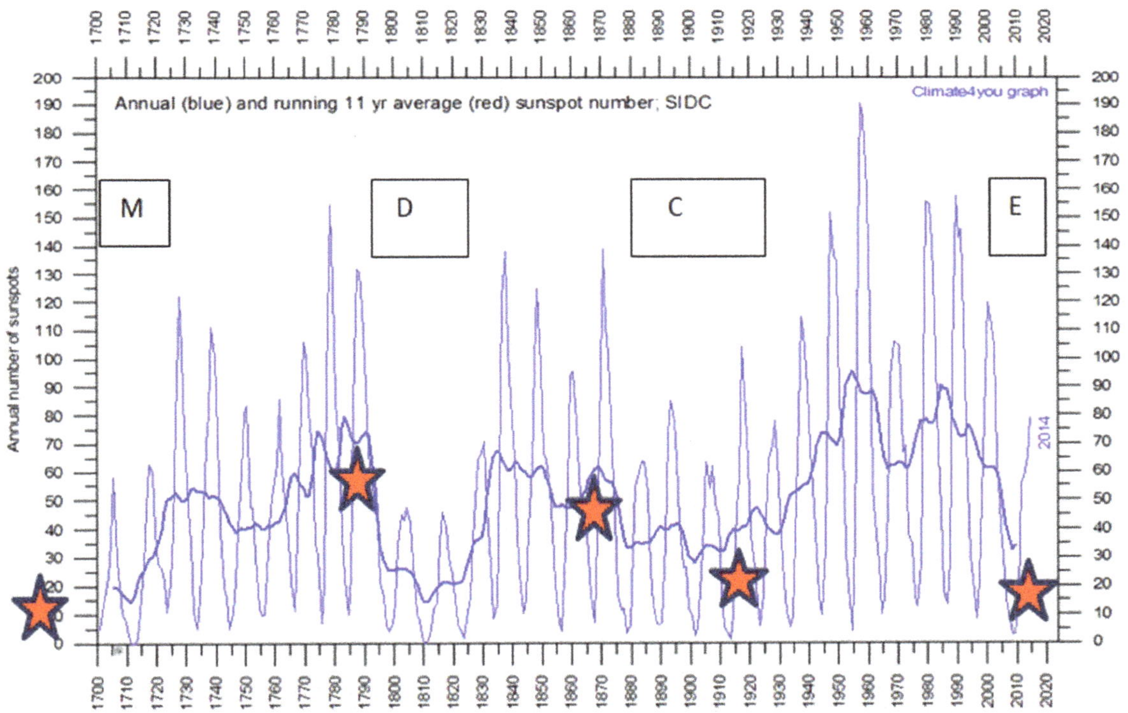

Figure 40. Comparison of Puerto Rican earthquakes vs. solar activity as measured by four hundred years of sunspots. The light blue lines are the 11-year solar cycles and the dark blue lines represent the smoother average of these many cycles. Red stars are, *left to right*, the quakes of 1670 (off the chart), 1787, 1867, 1918, and the 2014 M6.4 quake during the start of the Eddy minimum. The solar minimums are shown as the Maunder minimum (*M*), the Dalton minimum (*D*), the centennial minimum (*C*), and the Eddy minimum (*E*). Source: J. Casey, Dr. Ole Humlum, USGS.

In this chart we have the five largest earthquakes to strike Puerto Rico since the first in 1670, a large destructive quake though with no assigned magnitude by the USGS. That quake is placed off the chart in the bottom left. Going left to right from the 1670 event during the bottom of the Maunder minimum, we have the 1787 M8.0 quake, followed later by the 1867 M7.5 quake, the M7.5 1918 quake, and the most recent M 6.4 quake on January 13, 2014.

Again, much can be discerned from such an historical examination earthquake and sunspot records:

1. Four of the five (80%) of these catastrophic quakes took place at the bottom of a solar minimum (1670, 1918) or during a prolonged drop in solar activity (1867, 2014). The 1867 event was just after the multicycle pause of 1880 and at the same time was at a solar low point between 11-year solar cycles when sunspot counts were near zero.

2. Only one of five (20%), the 1787 quake, took place during a relative growth in solar activity as measured by sunspots. However, it happened only six years from the steep sunspot drop off into the Dalton minimum and just after the peak of sunspot growth up to 1784. It also took place during the low solar levels between two adjacent 11-year solar cycles. This one could be debatable.

3. The solar hibernation of the Dalton minimum saw no major earthquakes in Puerto Rico though in 2014, we have recently seen the first during the just started Eddy minimum. This quake took place at the peak of the 11-year solar cycle for cycle 24 and is in keeping with the Choi-Maslov relationship that suggests that a small spike of activity at the peak of the cycle is to be expected. The same type of event was observed in Alaska as mentioned in the previous chapter.

4. There had been no major earthquake in Puerto Rico for over 100 years until the January 2014 quake.

5. The average time interval between the five major quakes is 86 years. The four quakes before the 2014 event had an average interval of 83 years. The 2014 struck 86 years after the 1918 event, only 3 years after the expected date.

Here is the time frame analysis of major quakes to compared sunspot low points:

Year	Associated Low-Sunspot Year	Difference in Years	Comparative Year from 2031
1670	1670	0	2031
1784	1810	-26	2005
1867	1900	-33	1998
1918	1910	+8	2039
2014	2031	-17	2014

The absolute average of the four differences to the next solar minimum low point is 16.8 years, albeit with a large standard deviation. The associated time frame range of the five comparative years relative to the next major quake and to 2031 is 1998 to 2039.

The 2014 quake took place during the start of the Eddy minimum which is a long decline expected to last another 17 years from 2014 prior to reaching bottom in 2031. The 2014 quake took place at roughly the same average sunspot level as was seen in 1910, the centennial minimum.

Like California and Alaska, Puerto Rico is also subject to crustal energy flows which can be used to identify future high-risk periods. The status of the earthquake threat and underlying geology is explained in more detail in appendix 2 with a paper by Dr. Choi and Dr. Tsunoda. Some elements of which are covered in this section. From that paper we have the following revealing chart.

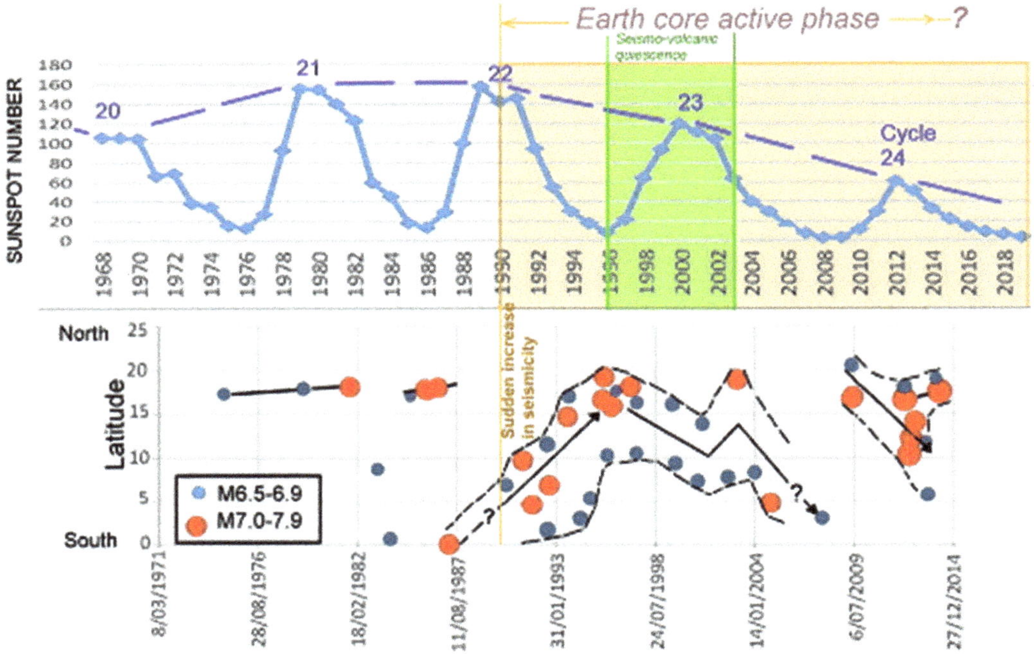

Figure 41. Solar cycles and earthquake propagation trend for Latin America. Note a general trend that earthquakes move northward when the solar cycle is in decline, but southward when the solar cycle in rise except for the period from 2005 to 2009 for which no data are available. A sudden increase in seismic activity from 1990 coincides with the start of declining period of a longer cycle obtained by tying the peaks of 11-year cycle. Here we see that with cycle 24 at about the latitude of Puerto Rico (~18 degrees N) we have an increase in large-earthquake activity. Sources: Choi et al., 2014a. Seismo-volcanic quiescence cited from Choi (2010) and Tsunoda et al. (2013) and the Earth core active phase from Choi and Maslov (2010).

What general conclusions with regard to future major Puerto Rican earthquakes may be drawn?

Before we do so, it should be mentioned that for the better part of a year now, the island has seen an record-setting swarm of moderate-size constant quakes, most of which are in the M4.0 to M5.0 range off the northeast coast of the island. This swarm may be one of the longest at that magnitude level in decades for any area of the globe. Our assessment is that the record-setting swarm could be due to one of the following reasons:

1. Development of an undersea volcano.

2. Precursor activity which may lead to a major earthquake.

3. A long-term earthquake in progress where energy is being released over a long period of time rather than through a single large quake. This possible slow earthquake scenario was also observed by researchers who studied the Cascadia Subduction Zone along Vancouver Island.

Without the benefit of any hard data, that is our best shot at what's going on with the Puerto Rican earthquake swarm. That aside, we believe the following is the picture for Puerto Rico's earthquake future:

1. The major earthquake for Puerto Rico that hit January 13, 2014 was close to the expected year based on the average of the previous four quakes. It also took place appropriate to a centennial minimum cycle quake schedule.

2. A question that comes out of the 2014 quake is whether it was the next and only large earthquake scheduled to hit Puerto Rico during the Eddy minimum. The answer is no. The solar hibernation associated with the year 2031 brings added potential for yet another quake, possibly of the category of the 1918 M7.5, one which would more than 32 times more powerful than the 2014 quake. The planet is now at a sunspot count that is similar to the centennial minimum when the 1918 quake struck. With that comparison, it would seem that the 2014 quake was on schedule relative to the count of sunspots. The process for continued decline in solar activity toward a deep solar hibernation is far from over as the Sun's energy output continues to decline as we approach 2031. The M6.4 quake is thus a wakeup call and not the end of the story for catastrophic quakes in Puerto Rico's future.

3. The 80% occurrence rate for past catastrophic earthquakes during solar declines or deep solar minimums indicates that another large and possibly catastrophic quake could take place during the ongoing Eddy minimum.

4. We believe the time frame for this next destructive quake for Puerto Rico as compared to past quakes versus sunspot low points will be during the period of 1998 to 2039. Yes,

> even with the 2014 quake, this means that, as of 2017, we are roughly 50% into the high-risk period for the next catastrophic earthquake.

The catastrophic earthquake future for Puerto Rico over the next ten to fifteen years is still developing.

Hawaii

Much like our review of Alaska, the unique geology of Hawaii is unlike any other U.S. state, including Alaska. We expected to see yet another repeat of the strong correlation between Hawaiian earthquakes and solar activity like the rest of the high-risk zones we have previously explored herein. The thought of a state essentially made up of volcanoes did create a small lingering doubt, however.

What we found was far better than a strong correlation. It was a strong anti-correlation! This island chain was born of a geological hotspot and is a living, breathing example of the dynamic forces that have shaped the Earth. The volcanic origin of these islands still dominate any assessment of earthquake risks there since, in effect, Hawaii is essentially a chain of volcanoes going back millions of years. It is not influenced by tectonic faults or similar earthquake-generating geophysical processes that we have looked at in the NMSZ or California, for example. It is a one-of-a kind U.S. state in its own geological category.

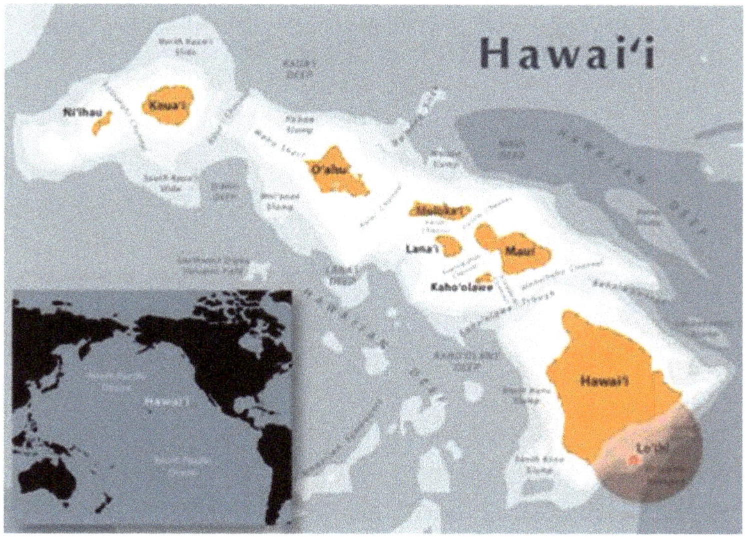

Figure 42. Map of Hawaii. Source: Wikipedia, the Free Encyclopedia. October 28, 2016, from https://en.wikipedia.org/w/index.php?title=Hawaii&oldid=746575421

Here is the brief history of Hawaiian earthquakes and tsunamis to strike the islands up to 1972 from the USGS:

> Much of the early record of Hawaiian earthquakes comes from the diary of Mrs. Sarah J. Lyman, a missionary's wife at Hilo, on the Big Island of Hawaii. Mrs. Lyman began her account in 1833 and continued it until her death in 1885; this record was then continued for eleven more years by her descendants. About four or five earthquakes per year were reported.
>
> On February 19, 1834, a strong shock threw down stone walls, stopped clocks, upset bottles, and sloshed milk out of half-full pans. Standing and walking were rendered difficult. A similar earthquake occurred on December 12, 1838. No volcanic activity was noted for either event.
>
> On March 27, 1868, whaling ships at Kawaihae on the west coast of Hawaii observed dense clouds of smoke rising from Mauna Loa's crater, Mokuaweoweo, to a height of several miles and reflecting the bright light from the lava pit. Slight shocks were

felt at Kona on the west coast and Kau on the flanks of the volcano. On the 28th, lava broke out on the southwest flank and created a 15-mile flow to the sea. Over 300 strong shocks were felt at Kau and 50 to 60 were felt at Kona. At Kilauea the surface of the ground quivered for days with frequent vigorous shocks that caused lamps, crockery, and chairs to spin around as if animated. One shock resembled that of a cannon projectile striking the ground under the proprietor's bed, causing him to flee, according to the narrative published by C. H. Hitchcock in the *Bulletin of the Seismological Society of America* in 1912. Between March 28, 1868, and April 11, over 2000 distinct shocks were felt at Kona.

The main shocks struck on April 2, at 4:00 p.m., and again on April 4 at 12:30 a.m. A magnitude of 7 3/4 was estimated for this earthquake (by Augustine Furumoto in his February 1966 article on the Seismicity of Hawaii in the *Bulletin of the Seismological Society of America*) based on the extent of intensity reports. Instrumental recordings, the usual basis for computing magnitudes, were not available at this early date. The shock was felt throughout the islands as far as Niihau some 350 miles away. The ground rolled like a ship at sea and many walls tumbled down. A landslide three miles long and thirty feet thick swept down the hill carrying trees, animals, and men. Thirty-one people and thousands of cattle, sheep, horses, and goats were killed in the one slide. A seawave struck the coast from Hilo to South Cape, being most destructive at Keauhou, Puna, and Honuapo; 180 houses were washed away, and 62 lives were lost to the wave alone. A 10-foot-high wave carried wreckage inland 800 feet. Not a house survived at Honuapo. A stone church and other buildings were destroyed at Punaluu. Maximum wave heights were 65 feet, the highest observed on Hawaii to date. (More on this earthquake.)

An intense earthquake occurred on January 22, 1938, with a magnitude on 6 3/4 and a maximum intensity VIII on Mauna Loa. The epicenter, located under the ocean about 40 miles east of the island of Molokai, was about as far north as earthquakes occur in the Hawaiian chain. On Maui there was general panic with people rushing from theaters. Flashing lights were reported by many. Landslides blocked roads and cut water pipes. Several reservoirs and water tanks were damaged. A chimney fell and a transformer was thrown down at Hana. Windows were broken and walls were cracked at Kula.

It was felt widely on the other islands with some damage on Molokai (pipes broken), Lanai (bottles thrown from shelves), Oahu (organ pipes out of sockets at Honolulu and the seismograph at the University was dismantled), and Hawaii (dishes broken, some chandeliers fell). The earthquake was distinctly felt by two ships at sea.

A severe earthquake occurred on August 21, 1951, and had a maximum intensity of IX and a magnitude of 6.9. Scores of homes were wrecked or damaged on the Kona coast on the west side of Hawaii. Rocks fell from cliffs, causing a 12-foot wave. A landslide covered the famed PaliKapu o Keoua burial grounds of Hawaiian royalty. Cracks six inches wide opened on the coastal highway. Walls of churches were thrown down in Hookena and houses moved from their foundations at Napoopoo and Kealakekua. Telephone service was out through most of the area. The collapsing of water tanks along the dry Kona coast faced with a two-month dry season made it necessary to truck water from Hilo.

Scores of small earthquakes are reported felt each year.

Hawaii is also exposed to another earthquake threat. In addition to the tectonic and volcanic local earthquakes it is a frequent victim of tsunamis from distant earthquakes. The *Catalogue of Tsunamis in the Hawaiian Islands* by George Pararas-Carayannis list 85 tsunamis since the earliest reported in 1813 or 1814, of which 15 have caused significant damage. Only four of these, including the 1868 earthquake and tsunami described above, have originated near Hawaii. Most have originated in the northwest Pacific and near South American coasts.

In 1837 an earthquake in Chile sent waves 20 feet high against Hilo, Hawaii. Initially the sea receded and several were drowned by the returning wave while they were attempting to collect fish stranded on the exposed sea bottom. In all, 62 people were killed and over a hundred homes were destroyed.

The most destructive tsunami in Hawaii occurred on April 1, 1946, following an Aleutian Islands earthquake. Waves 55 feet high, crest to trough, struck the northeast coast of Hawaii. At Hilo, 173 were killed, 163 injured, 488 buildings were demolished and 936 more were damaged. Damage was estimated at $25 million. The waterfront was washed out and breakwater and wharves badly damaged. This tragic loss of life prompted the formation of the Tsunami Warning System so that Hawaii and the countries bordering the Pacific would never again be surprised by the large destructive waves. (Abridged from the *Earthquake Information Bulletin*, volume 4, number 1, January–February 1972.)

Just as we did with the other earthquake-prone states, a solar activity chart versus earthquake history was put together for Hawaii. Here it is:

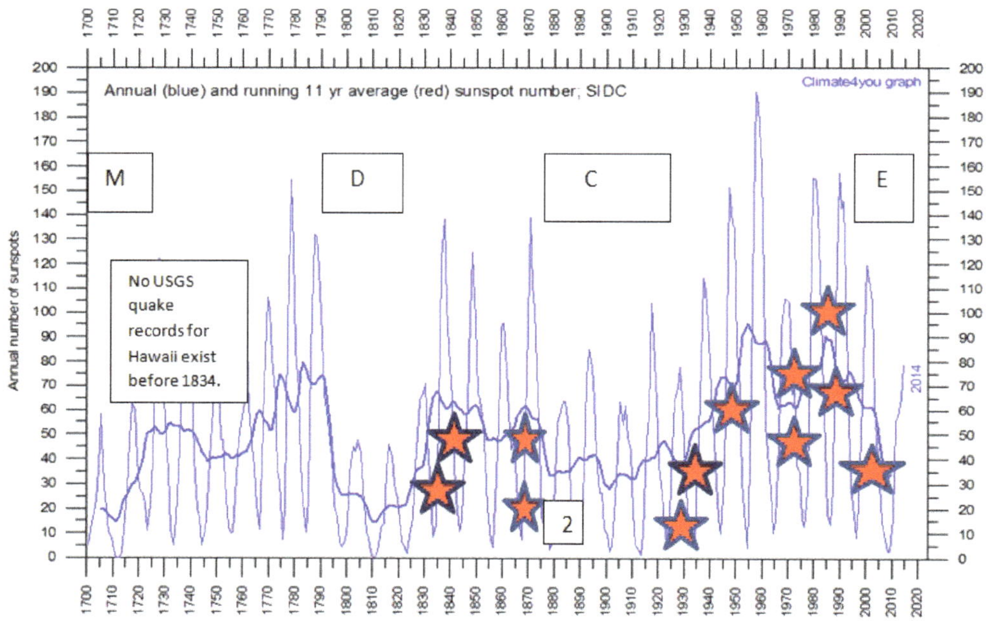

Figure 43. Hawaiian earthquakes compared to solar activity. Stars on the chart represent individual earthquakes except in the year 1868 when two occurred as taken from the table of earthquakes compiled by J. Casey. Sources: J. Casey, "Hawaiian Earthquake History," Dr. Ole Humlum, USGS.

It was a challenge finding a relatively complete set of recorded major (M6.0+) earthquakes for Hawaii. Here is what was found from the several cited sources:

Hawaiian Earthquake History
Compiled by J.Casey

Number	Date	Location	MM	Deaths	Description	Source*
1	09/19/1834	Oahu	UNK		Stone walls toppled	1
2	12/12/1838	Oahu	UNK		Similar to 1834	1
3	03/28/1868	South Hawaii	7.0	0	Extensive in Southern Hawaii	
4	04/02/1868	South Hawaii	7.9	81	>100 houses destroyed, tsunami	
5	02/ XX/1871	Lanai	6.9	UNK	Extensive in Kona	2
6	10/05/1929	Hualalai	6.5	0	Cracked water catchment and oil tanks and broken water pipelines, buildings damaged	
7	01/ XX/1938	Maui	6.9	UNK	Extensive in Kona	2
8	08/21/1951	Kona	6.9	0	Extensive in Hilo	
9	04/26/1973	North of Hilo	6.2	0	Extensive in Hilo, $5.6M	
10	11/29/1975	Kalapana	7.7	2	Extensive in southern Hawaii, $4.1M	
11	11/16/1983	Ka'ōiki	6.7	0	Extensive in southern Hawaii, >$6M	
12	06/25/1989	Kalapana	6.2	0	Southeast Hawaii almost $1M	
13	10/15/2006	Kīholo Bay	6.7	0	Northwest Hawaii, >$100M	

The table above was compiled from the following sources:

1. USGS Hawaiian earthquake history. Quakes listed assumed M6.0+
2. Article titled "Reminders for Hawaii residents to prepare for damaging earthquakes," *West Hawaii Today*, published April 18, 2016

3. "Destructive Earthquakes in Hawaii County Since 1868," http://hvo.wr.usgs.gov/earthquakes/destruct/

**All* are from source 3 except as noted in the table.

****

A plot of the thirteen major Hawaiian earthquakes found the following:

1. Of the thirteen major quakes cited, a significant number (9) occurred during a growing period of sunspot or solar activity. That represents ~69%.

2. Only four earthquakes—1929, 1973, 1975, and 2006—took place during a decline of sunspots, an MCP, or a prolonged solar minimum. That represents ~31%.

3. A total of 58 years elapsed (1871–1929) representing the long centennial minimum period without any major earthquakes on the Hawaiian Islands.

4. There appears to be a trend of earthquake pairs or triplets of major quakes. This has been happening within a period of 2 years to 9 years of each other covering the 172 years of earthquake history in the compiled table. This shows that only a narrow 24 years out of the entire earthquake history of 172 years involves 11 of 13 major earthquakes or 85% of all Hawaii's earthquakes. Here are those pairs and total years separation:

1834, 1838 – Four years separation

1868, 1868, 1871 – Three years

1929, 1938 – Nine years

1973, 1975 – Two years

1983, 1989 – Six years

5. Though there is a lack of data during the Maunder and Dalton minimums, it appears that:

> *The Hawaiian Islands react by producing major earthquakes during periods of increasing solar activity, i.e., periods of increasing sunspots and associated global warming. This is a strong anti-correlation with the cold epochs implied by the RC theory and the comparable earthquake histories for all other U.S. states.*

Therefore, what we can say about the possibility of future quakes in Hawaii is that the next large quakes should be around the time the next predicted growth in solar activity and concurrent global warming is predicted by the RC theory. This should be in the late 2040s or thereafter. Not including the 2006 earthquake, there remains a small roughly one-in-four chance of a major earthquake during the next solar minimum or the Eddy minimum.

On the surface, this finding of a reverse characteristic for hotspot-linked earthquakes may seem to contradict all other findings in this book. In fact, it does not. Here's why:

1. It indicates what has already been stated. The Hawaiian Islands earthquakes are volcanic in origin caused by a geological hotspot, i.e., volcanic activity and not traditional earthquakes originating from crustal faults, subduction zones, or continental separations. Out of the 50 U.S. states, Hawaii is truly different.

2. A negative or increasing trend of major earthquakes has the potential to be just as relevant to climate related trends associated with positive or decreasing earthquake trends. Further study of the phenomena is needed.

3. For reasons not yet understood, global hotspots like Hawaii may represent an inverse or one-off relationship to far more numerous causative factors that produce catastrophic earthquakes around the world.

Rather than use this outcome to compare with the findings of earthquake risks for other U.S. high-risk zones, perhaps a better approach to understanding Hawaii and its CGE threats to the people there may be to compare it to other geological hotspots. That would then be an apples-to-apples comparison. In other words, to compare Hawaii to the other U.S. high-risk zones is more like an apples-to-oranges situation.

One of the core purposes of basic research is discovery of something highly relevant and yet something that adds, if not runs contrary, to the hypothesis under consideration. Hawaii has supplied us a new geophysical phenomena not only to explore but also one that in fact may link solar activity by another process to geophysical activity on Earth.

What was expected in the case of Hawaiian earthquakes became something unexpected. We are delighted with this final geophysical analysis of a U.S. state in this text. In a purely random organization of subject matter when the outline of this book was constructed, the Hawaiian analysis, the last one planned, has turned out to be one with great promise for future research.

Chapter 7

The Greatest Threat: Multiple Catastrophes

Go forth to meet the shadowy Future without fear and with a manly heart.
—Henry Wadsworth Longfellow

Most of us have been through a natural disaster or know someone who has. It may have been a cat 3 or larger hurricane like the 1992 Hurricane Andrew, been affected by the direct or downwind effects of the Mt. St. Helens eruption of May 18, 1980, survived the 1964 Alaskan M9.2 quake, experienced one or more serious M6.0+ or greater earthquakes in California, or any number of devastating tornados that hit the United States every year. Even with these events and their resultant damage and loss of life, they were essentially isolated, and except for those communities in harm's way, the rest of the United States was able to go on as normal so much so that these events were pushed back into memory within months or a few years at most. Unless they struck a major metropolitan area like the Northridge 1994 or the 1989 Oakland quake or a Katrina-size hurricane, they fortunately did not take a lasting toll on most Americans in our large diverse country.

It is doubtful that a typical man-on-the-street interview would find many, especially among the under-30 crowd, who could remember what the words *Mt. St. Helens* or *Andrew* or *Katrina* mean in terms of natural disasters. In any case, they still do not compare with the truly catastrophic quakes that have devastated other areas of the globe, especially China and others where tens and hundreds of thousands of people have been killed by monster earthquakes. See appendix 3.

All that is about to change as you will see in this chapter because *the greatest geophysical threat we may have to face as a nation and as a planet is the likely occurrence of multiple catastrophic earthquakes within a relatively short period of time in the next decade or two. If history is our guide, that is exactly what will happen.*

A review of what we have already discovered about the largest earthquakes by region points the way to understanding how serious and just how real this nationwide threat is. In this chapter, we will focus on this greatest threat.

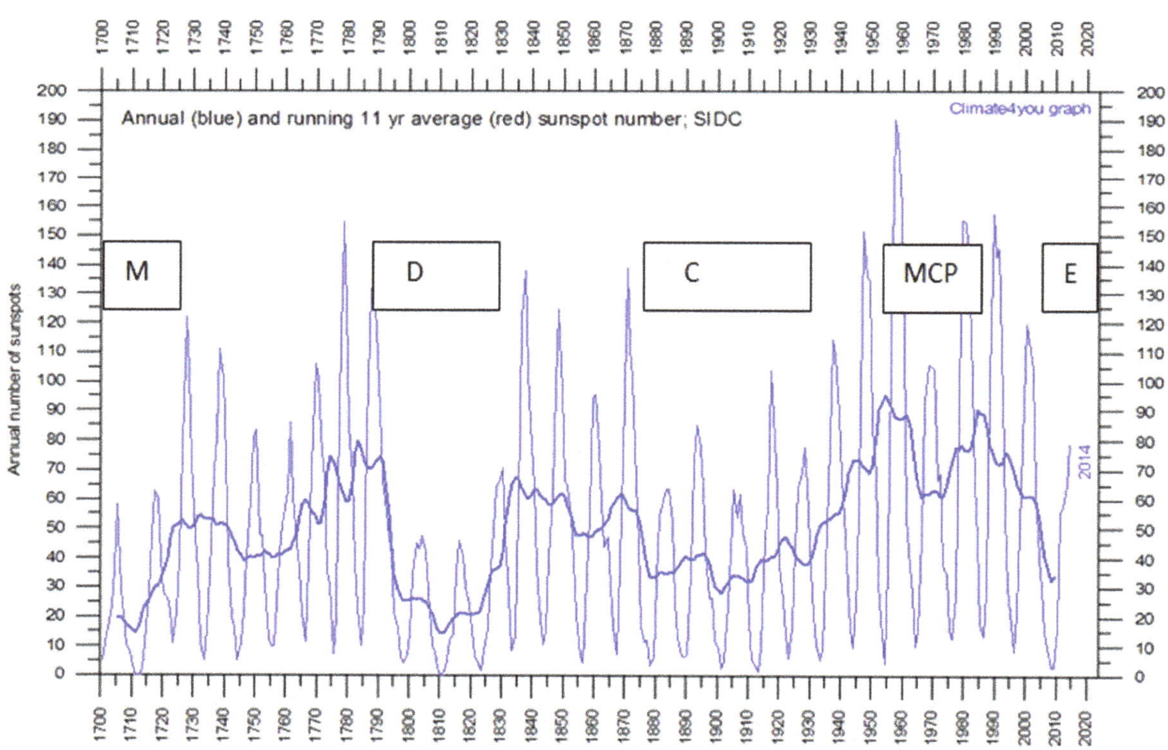

Figure 44. Solar activity measured by sunspots from 1700 to 2014. The light blues lines are the 11-year solar cycles' sunspot counts. The dark blue line is the smoothed average of the sunspot counts. *M, D, C, MCP,* and *E* represent the Maunder minimum, Dalton minimum, centennial minimum, multicycle pause of the 1970s, and Eddy minimum respectively. Source: J. Casey, Dr. Ole Humlum.

The chart above is for reference as we examine below what we have learned thus far about when the largest U.S. earthquakes strike. We have already established from those previous assessments that there is a strong correlation between solar minimums and solar hibernations (e.g. Maunder and Dalton minimums). Some of the big earthquakes in the United States have also happened even during lesser solar activity lows like the centennial minimum and the multicycle pause.

Because we have had never had to live through such combined natural disasters before, nor our parents, or their parents or their parents before them, we are thus psychologically and culturally insulated from the prospect of such events and the veritable waves of calamites they might pose. They haven't happened on this scale before to anyone's immediate knowledge, so why should we be worried now, right?

Let's take a look once more, at the list of quakes that happened for each of our high-risk areas during the low solar activity periods, but this time categorized by these minimums:

The Maunder Minimum

1. 1670 Puerto Rico Mag. UNK
2. 1700 Cascadia Subduction Zone (CSZ) M9.0
3. 1695 NMSZ Approximately M7.0–M8.0

The Dalton Minimum

1. 1811–1812 NMSZ M7.3, M7.5, and (M7.5-M8.0)
2. 1812 California M6.9
3. 1812 California M7.1
4. 1817 South Carolina M4.8

The Centennial Minimum

1.	1895	NMSZ	M6.8
2.	1886	South Carolina	M7.3
3.	1899	California	M6.7
4.	1899	Alaska	M8.2
5.	1899	Alaska	M8.6
6.	1900	Alaska	M7.9
7.	1906	California	M7.8
8.	1912	South Carolina	M 4.8
9.	1914	South Carolina	M4.2
10.	1918	California	M6.7
11.	1918	Puerto Rico	M7.5
12.	1923	California	M7.2
13.	1925	California	M6.8
14.	1927	California	M7.3

The Multicycle Pause

1.	1964	Alaska	M9.2
2.	1965	Alaska	M8.7

3.	1971	California	M6.1–M6.7
4.	1974	South Carolina	M4.3

The Eddy Minimum

1.	2014	Alaska	M7.9
2.	2014	Puerto Rico	M6.4

What does the table above reveal? It gives us a clear indication of what we can expect during the Eddy minimum if earthquake history matches predicted solar cycle behavior. Here is the most likely scenario for our immediate geophysical future (2017 to 2043) under the assumption that a Dalton-class minimum approximates the coming Eddy minimum plus only 25% of the centennial and modern multicycle pause earthquakes repeat. This is a conservative number of additional quakes in view of the increasing rate of earthquakes globally, and the already-identified steep drop in solar activity off of solar cycle 24 and an expected Eddy minimum lasting at least 30 years. This estimate may be adjusted upward should the predictions of a Maunder-class minimum come about instead of another Dalton-class minimum.

Here is what we should expect to see in general during the Eddy minimum:

1. We should plan on a period of unparalleled geophysical destruction not seen in this country since its founding. This will include an unbelievable number of catastrophic earthquakes striking several high-risk zones of the United States either within the same year or within a small number of years from each other. Planning should include worst-case scenarios that realize three to five M7.0 or larger earthquakes in total, including California and the NMSZ within six to ten years of each other.

2. We should expect to see another catastrophic earthquake series (three catastrophic earthquakes of M7.3 to M8.0) over several months to strike the NMSZ, causing widespread death and destruction within the eight central Mississippi valley states. Damage to the NMSZ states will be at least that estimated by FEMA at the $300 billion level, and we believe closer to the $600 billion level borne out by their study. In addition, there will likely be significant collateral damage to the northeastern states and the USA economy,

significantly adding to the total property and economic and personal losses above that in the eight NMSZ states.

3. We predict that a segment C rupture of the Cascadia Subduction Zone will occur during the cited risk period of 2017 to 2043.

4. California should see at least two and possibly four catastrophic quakes in the range of M6.9 to M7.8 during the cited period, ranging from San Francisco to Baja. This will result in widespread damage and loss of life throughout the state and substantial additional damage and loss of life through subsequent social and economic destabilization.

5. Alaska has already started the Eddy minimum record of catastrophic earthquakes with the Rat Island M7.9 quake of 2014. This is no surprise given the long history of catastrophic earthquakes in that state. We should expect to see at least two more in the range of M8.2 to M9.0 during their high-risk period.

6. South Carolina should expect to have two more moderate quakes in the M4.2 to M5.0 range and at least one catastrophic quake at the M7.0+ level.

7. Puerto Rico will likely have a catastrophic earthquake at the M7.0 to M7.5 level, resulting in substantial loss of life and devastating property and economic losses along with long-lasting social disruption. The predicted event in Puerto Rico could occur at any time since they are well within their predicted high-risk period.

Impacts on the U.S. Government

The predicted catastrophic earthquakes that are documented in this text unfortunately will occur in rapid succession, causing an historic tragedy nationwide. This scenario is based on probabilities of near or over 80% in all high-risk earthquake zones that have been examined. Hence, there is a roughly 80% chance that this set of predictions above will unfold for the nation as a whole to the degree and duration identified by the research. This will produce long-lasting adverse effects, challenging the U.S. government to meet its internal and external obligations across the board in all areas and agencies of the federal government.

Here are the possible nationwide results that may come from the postulated scenario and the itemized damaged done within the various high-risk states:

1. The NMSZ quakes alone are likely to cause years of national economic and social distress. The expectation for economic losses to the central Mississippi valley states and the northeastern U.S. states are going to be horrendous. We believe the FEMA estimate of damage from the predicted NMSZ rupture is understated. It does not, for example, include any collateral damage from a complete loss of oil and gas supplies for many months that provide heat and energy for millions of homes and business in the northeastern United States that are currently delivered through pipelines that traverse the NMSZ.

2. The U.S. GDP will likely receive a major hit from this quake accordingly. With the nation already at over $19 trillion in direct debt, we believe there is a real prospect for a long-lasting depression following an NMSZ event which could be a $600-billion calamity and would seriously impact the federal government's ability to honor its financial obligations for those currently covered by any form of federal entitlement payments, much less meet foreign and investor debt repayments. Treasury receipts are likely to decline significantly during the depression while at the same time obligations and emergency aid requirements to rebuild entire areas of the U.S. infrastructure would be growing.

3. This situation is compounded by the tragedies that are expected in California and other states that are already near bankruptcy. We expect California will be unable to meet its obligations in view of the coming catastrophic damage, property losses, and economic losses. The U.S. government which will be under stress at the same time may be unable or unwilling to bail out California or other states heavily impacted by the predicted years of devastating earthquakes.

4. It should be no surprise that we will see substantial criminal activity in the heavily damaged areas, especially because of long-term loss of basic needs like jobs, food, clean water, and power for heating during frigid winter months or cooling during hot summers.

5. The new cold climate that is associated with the predicted rash of major earthquakes also during roughly the same time frames as the earthquakes is well documented within the solar physics and climatology communities. The forecast for a Dalton-class solar minimum or solar hibernation includes major crop losses from new record-setting cold climate ill effects. The forecast by Russian climatologists and others for another Little Ice

Age will create a new global economic and political landscape with a likely outbreak of at least regional warfare.

6. Though not covered specifically in this book, prior research has also established that catastrophic volcanic eruptions are also highly likely during the Eddy minimum. Just like major quakes, major volcanic eruptions coincide with these solar hibernations. These volcanic effects are expected to be global in nature though with no as-yet-quantifiable impact on the USA and its future crop yields and energy requirements. However, during the last solar hibernation, the Mt. Tambora volcano in Indonesia exploded and created a veil of dust and SO_2 that went around the world and contributed to what has been called the year without a summer for 1816. During that period, thousands of Americans in New England starved and froze to death, and snow and freezing temperatures were observed in August 1816.

7. The total set of adverse geophysical effects to be seen during the coming two decades has the potential to test the country's survival to a degree that has not been seen since World War II.

****

Because multiple large-scale natural catastrophes that occur close together do so only on the longer cycles of the Sun measured in hundreds of years or longer, we simply don't take the threat of them occurring during our lifetime as serious concerns. Such historical events regarding climate and geophysical catastrophes and their impact on human civilization are not taught in our school systems or covered in any media of our day and are certainly now being ignored by our government.

To our knowledge, there is not one college course taught today that discusses how the Sun causes climate change and the Sun's involvement with major earthquakes (or volcanic eruptions). Why is this sad situation the case? Because it is politically unacceptable to consider this new geophysical reality or that there could be any theory of climate change more important or more reliable than the greenhouse gas theory.

It will remain so until scientific integrity is truly restored within the U.S. government and our scientific institutions.

Chapter 8

Preparing for the Great Upheaval

The Grasshopper having sung
All the summer long,
Found herself lacking food
When the North Wind began its song.
Not a single little piece
Of fly or grub did she have to eat.

—Jean de La Fontaine,

From the poem "The Ant and The Grasshopper"

Is preparation for the coming geophysical upheaval even possible? How one prepares will be a function of resources, knowledge, and an understanding of past events, especially those regarding how large numbers of people react after they have been deprived of life's essentials.

This chapter is not intended to be a specific guide on survival techniques or a thorough treatment of the extensive world of preppers, food storage, how to hunt or fish or raise one's own crops, etc. There are numerous manuscripts and books available for those who want to dig in deep into these subjects. This chapter is intended, however, to provide some fundamental recommendations on how to begin one's preparations for the specific earthquake threats discussed in this book.

There are several issues to consider based on a general scenario for getting safely through the catastrophic earthquakes that are coming. Most important above all is that to minimize the ill effects of the difficult years ahead, you must prepare as if no one will come to your aid. Sadly, if you are in the most heavily damaged earthquake zones, you must also prepare, in a worst-case scenario, for others who may come to beg for or physically take what you have set aside for your or your family's survival!

Survival of any natural calamity is always improved when people are independent of the government or others for their own survival. The situation in the USA, however, could not be worse from a perspective of personal responsibility and self-reliance.

> Not only is the vast majority of Americans dependent on large utility systems that supply water and power, but there are also now tens of millions of Americans totally dependent on the government for food and security. In addition, our current population has a growing number of people over age 65 made up of baby boomer retirees, a growing number of whom who are under family or assisted-living care. This large segment of the population is partially or completely reliant on friends, family, or caregivers for their day-to-day needs.
>
> There are also tens of millions more who, for political reasons, will not bend to the notion that mankind does not control the Earth's climate or climate-induced geophysical effects. They are the ones who will refuse to prepare for the difficulties that are coming and are content to believe that the government will care for them. These government dependents and unprepared people are the grasshoppers of Jean de La Fontaine's poem. Being an ant and adopting a more prepared, secure way of life that is founded on self-reliance and self-determination and not wanting to fall victim to life's unknowns is of course preferred. The grasshoppers are not comfortable with this thinking even if the best advice tells them it may save their life. They are more content reacting to what they want the future to be rather than what reality tells them it will be. It is living in a dream world. This may tragically lead to them being among the first on the list of dead at city morgues after natural catastrophes of historic scale come one after another for the next twenty years.

One of the riskiest aspects of preparing for the predicted catastrophic geophysical events (CGE) of the next twenty years is assuming that local, state, and federal governments will quickly come to your aid. The sooner you dismiss this unlikely possibility, the more viable your post-CGE survival planning will be.

Accordingly, this chapter of this book does not involve any guaranteed government assistance at any level at any stage before, during, or after a CGE. In fact, there is reason to believe certain government elements may be just as dangerous to your family's survival as the criminals on the street.

How the Government May Turn against You after a Catastrophic Earthquake

This chapter includes recommendations for the government, but as is typical of large self-adulating, self-protecting government establishments, they will neither seriously consider nor act on these recommendations—that is their proven historical response until tragedy strikes. Remember that after Hurricane Katrina hit in August 2005, some members of the New Orleans Police Department resorted to committing crimes instead of protecting innocent people from criminals.[15] FEMA said they had learned from the Katrina fiasco, yet recently when tens of thousands were forced to flee from the historic flooding of Louisiana in August 2016, the response by FEMA was described as pitiful by Congressman Mica (R-FL) on the U.S. House committee that oversees FEMA.[16] One of the things I am most concerned about is the threat posed by state and federal government agencies and what it may take from you if you are among the best prepared to protect your family.

In a worst-case situation, we should not be surprised that local and state government officials may require you to deliver your survival stores to the government directly who will distribute them as they see fit. They may rely on some obscure antihoarding law or local disaster ordinance to force you to comply or in order to quell a complaint from hungry neighbors. Stocking up on essentials for survival is not hoarding if done before a disaster strikes. It's smart planning. Authorities or your neighbors may not see it that way though. They may require you to spread your food and water among the unprepared people, the grasshoppers on your street, the same people who probably laughed at you for preparing before the earthquake struck! Yet instead of them showing gratitude as their emergency provider of critical life-sustaining food and water or power, your neighbors and the government may all of a sudden call you a hoarder and an evil, selfish person—justification for their brazen theft.

You should be prepared for your food to be confiscated by the government to ensure food equality or food justice because of politicians' verbal demands or laws passed by post-earthquake emergency acts of the government at the federal, state, or local level. These are expected terms the government may use to justify taking what you have to give to others who irresponsibly refused to prepare themselves even though they had ample warning from competent authorities or experts long before a devastating earthquake struck. That is the nature of big government, especially one driven by socialist ideology. After a regional or statewide M8.0+ catastrophic earthquake strikes, the well-known phrase "it takes a village" could mean the village is coming to take from you!

The General CGE Scenario

In consideration of how to prepare for a CGE, we might be well served to define what type of CGE we are planning for and what the community environment might be like after one strikes. For the purposes of this exercise, we will assume this CGE is an earthquake of magnitude 8.0 or larger. Volcanoes typically provide more early warning than earthquakes do, allowing much more preparation and thereby posing a lesser immediate threat and often allow evacuation from the danger zone. This was the case for Mt. St. Helens in 1980 and Mt. Pinatubo of 1991. St. Helens was a regional problem. Pinatubo, however, was such a huge eruption that it affected the Earth's climate for three years, causing a steep drop in global temperatures as it cast a veil of dust and sulphur dioxide high into the stratosphere.[17]

Another CGE, a tsunami, for example, along the northwest coast that affects Oregon and Washington State would be from a mega-thrust earthquake along the Cascadia Subduction Zone. It would cause horrific ill effects after the quake had done its damage. Still an M8.0 San Andreas quake would have no associated tsunami. With these kinds of considerations, we will focus in this chapter on the CGE being a single inland M8.0 quake (with no tsunami) in Alaska, the San Francisco Bay, San Andreas, or NMSZ, for example. A South Carolina or Puerto Rico quake of this magnitude is not likely based on historical trends. Even those living there would be better off if they followed the guidelines we recommend here for a larger M8.0 quake.

Each person living in a CGE threat zone should prepare according to the full set of threats that might come from one's particular zone as identified in the previous chapters in this book.

Possible Adverse Outcomes from an M8.0 Earthquake

a. First Responders

The normally quite-capable first responders, including police, fire fighters, EMTs, and other rescue personnel including local state and community emergency management agencies (EMA) would be overwhelmed in the first minutes after the catastrophic earthquake hits. If the 911 dispatch lines are still operating, they would be brought to a standstill because of the flood of tens of thousands of injuries and fires. These same first responders would have their own hands full trying to care for their own families who may be injured or in serious distress, possibly fighting their own fires or collapsed structures at home or work. Is a policeman, firefighter, EMT or hospital, or senior-care facility employee going to come in for work just like any normal day, risking harm to their families who are going hungry or are without water and power because everything in the grocery store was bought out or looted out overnight on the first day after a major regional earthquake?

b. Communications Systems

Within seconds to a few minutes after the quake strikes, the public communication lines would likely be severed except for some TV and radio stations with backup systems. Then the TV would be available only for those off-air TVs capable of receiving their transmissions since cable and satellite systems may be offline. The satellites in orbit of course are unaffected. It is their ground receiving and transmitting stations that would be knocked out and their underground signal and power lines that will be taken out. Only those ground stations with reliable backup power would be functioning. The general public may try to call 911 or listen to radio or TV stations, but it is possible the entire network of public communications, including home and business phone systems would be shut down if not irreparably damaged for months. Cell phone systems would be out of commission as well with towers lying on the ground or otherwise unable to operate. Tactical military-style FM and citizen shortwave (AM) and emergency radio systems may be the only ones operating. The next-door neighbor who has had the eyesore of an AM radio tower in his backyard for years may suddenly become your best friend.

c. Complete Loss of Power for Weeks or Months

A regionwide quake of M8.0 or larger could totally disable a multistate region's power grid for an extended period of time. Recovery from this scale of a CGE would not be like other local quakes where services can be restored in days or a few weeks at most. Here, we are talking about large power grid networks that would take months if not longer to return to full service. Large transformers and critical substations could not be quickly repaired even with available replacements from utility companies. Above a certain number of damaged critical power generation facilities and substation hardware, new units could take months to build and ship in. Some vital components of the grid are only available from overseas suppliers which again take months to build, ship, and install.

The loss of the national power grid or a substantial part of it could have long-lasting ramifications as Ted Koppel's book *Lights Out* (Crown Publishers, 2015) shows. In it he talks about a cyberattack or electromagnetic pulse (EMP) that brings down the grid. A regional-scale earthquake can have the same effect for a number of states except with an earthquake, the damage is physical, not just digital. The loss of major hard components of the grid through direct destruction by an earthquake could not be one where repairs and return-to-normal operations could be made in a short period of time. In referencing a group of experts who advised a national plan to protect the power grid, Koppel quotes a letter from the group:

> Under current conditions, timely reconstruction of the grid . . . if particular equipment is destroyed would be impossible; and according to government experts, would result in widespread outages for at least months to two years or more.[18]

If by some miracle, the region's grid were equipped with the kind of emergency shutdown system similar to that used in Japan and now proposed for Southern California, it might be able to salvage some elements of the grid. Those power systems and critical emergency systems equipped with such Japanese-style speed-of-light communications and deactivation capability might be able to safely shut down key elements, thus saving them from permanent damage. They would be taken offline in any case. The grid would still be down. Millions would still be without power for weeks or months if not much, much longer.

The advantage for these ShakeAlert systems comes during the period immediately after the earthquake starts where time is of the essence to later restoring essential services and full grid availability and shutting down high-speed people mover systems like commuter trains. These earthquake damage prevention systems like the $150-million system being considered for Southern California is at best able to provide "a few tens of seconds," i.e., 30 seconds' warning after the earthquake has already begun.[19] It is important to differentiate between an earthquake-prediction system, like that being developed at the IEVPC, and a ShakeAlert-type system. Whereas the IEVPC system is intended to warn of an earthquake days, weeks, or months in advance, ShakeAlert sends out a brief warning after the earthquake has already started!

A ShakeAlert would afford essentially no practical time for people to get out of buildings about to collapse or get off interstate or freeway bridges and overpasses about to crumble beneath them as currently planned. Those near the epicenter would receive effectively no warning time. Only those system participants far away from the epicenter would be able to benefit from such a system.

Our opinion of the ShakeAlert system as it is currently proposed in California is that it is an underscoped, pinch-penny version of what is needed in every high-risk zone in the USA, including along the West Coast, Alaska, the NMSZ, South Carolina, and Puerto Rico. The system should be expanded in scope and then fully funded as a national high priority well above the inadequate levels now being considered. There is far more than can be done with such a system in terms of saving lives than is now being considered in its limited application and design.

Unfortunately, the federal government has only given lip service to the concept and no meaningful funding, leaving California scientists, geologists, emergency management, contractors, and local leaders to scrape for every penny to field a system with a necessary national purpose.

The lack of meaningful federal support to the ShakeAlert system by the current administration in Washington will be paid for in human lives in the not-too-distant future.

For a modern society to sustain a complete loss of power over a large state or multistate area can be a tragedy of immense scale. The closest we have come is the power grid shutdown in the northeast on the August 13, 2003 Northeast Blackout which fortunately lasted only seven hours and did not strike in the middle of winter. Ultimately, the cause was found to be a failed alarm in one company's system that did not warn of a power overload. It triggered a cascade of failures of the neighboring grids that affected 10 million in Canada's Ontario province and 45 million in northeastern states.[20] Had this happened in the dead of winter or lasted longer, the region's social structure would have collapsed. The death toll and property destruction would have been record setting.

The situation described in this chapter is far, far worse. When the grid goes down and stays down for weeks or months, our modern world comes to an abrupt end all too quickly. Without power, we have no heat, no refrigerated food, only a day or two of gasoline before all the stations without backup generators run out of fuel or the ability to pump it. The lines to get fuel will be incredibly long and sometimes sites of violent confrontation. With no communications, no cell phones, no TV, no way to safely travel, and voila, all of a sudden we are back in the caveman days.

From 2004 to 2007, Florida and Atlantic and Gulf Coast states were hit with numerous powerful hurricanes. For those who went through these disasters, doing without power for days to weeks remains a vivid memory. Fortunately, the grid could be restored and friends and family who still had power could provide a brief respite from hurricane-season warm temperatures in their air-conditioned homes with plenty of light and operating refrigerators stocked with food. Even those hurricanes did not compare with the utter devastation of Hurricane Andrew that hit Homestead, Florida, on August 24, 1992 with 200-mph winds effectively leveling most of Homestead and surrounding areas just south of Miami. For those experiencing Florida's first land-falling hurricane in eleven years with Hurricane Hermine (September 1, 2016), the memories of the need for preparations without power are coming back. This story was again repeated with the long coastline path of destruction and power outages caused by Hurricane Matthew during the first week in October 2016. Massive flooding and localized power outages were the main impact in the USA. Haiti got the worst of it in terms of hundreds killed as all major newspapers covered.

Even with Matthew, in a week or two, most got their power back on. Once more, the general populace thought "Back to normal, no need to prepare."

The kinds of human behavior we saw back then during the period of our worst hurricanes from 2004 to 2007, i.e., fights at gas stations, crazy drivers at dead street lights, long lines at stores, were but a hint of the massive list of problems people will have to confront for the first time in their life and the resultant utter chaos we should expect to see during the CGEs we discuss in this book. *Upheaval!* is not just about geology, but also about psychology, human nature, and specifically how people react during desperate times. Challenges above the normal routine bring out the best and the worst in us.

These predicted catastrophic earthquakes can suddenly strike without warning and have one's life thrown back to the time 200 years ago—before electricity, when the most common form of transportation was not a bus, sedan, or a Ford F-150 truck, but a horse.

The power-outage story and hurricane histories recounted above makes one think of the difficult times ahead that will be compounded by the millions of our fellow citizens who are untested, uninformed, and unprepared. They will be the most vulnerable and the most shocked and, therefore, least likely of surviving unscathed in an extended period of time without food, water, and electricity. They will soon be tested on how they can deal with true upheaval in their lives. These periods of times when our world gets turned upside down will not just come and go after a few hours or days without power. The initial hours without powered electrical appliances and modern conveniences will go on for days, then weeks, and then into months or longer for those in the primary earthquake zones.

d. **Lack of Food and Water**

This is the most serious ill effect of a regional major earthquake. The typical grocery store has a three-day supply on the shelves and is dependent on truckers bringing in routine shipments on a regular schedule.[21] .After a power loss and without backup generators, the refrigerated items still in the stores would be the first to go through decay and spoilage. We should assume, however, that these stores will be emptied by frightened shoppers or the inevitable gangs of looters that will suddenly appear within the first 24 hours after the quake. It is unlikely those stores will see any resupply trucks any time soon.

What this scenario says is that unless one already has food and water stored in sufficient quantity, you won't be able to eat after the kitchen pantry is exhausted. Further, you should accept that there will be no food to find anywhere until social order is restored weeks or months later. If your friends and neighbors believe you have food stored for rough times, they will be the first at your door. Some of these former friends may be armed as the general situation deteriorates. Truckers will not be driving into a war zone to provide new stocks for store shelves with roads and bridges destroyed. Knowing they will be targeted on the road before they reach the store, some may be reluctant to proceed without an armed escort. Once your home power is out, you won't have anything in the refrigerator except rank food that will be unsafe to eat. You won't be able to save any frozen foods by cooking since the power will be out. A home barbeque grill will only get you so far before the propane tank is empty or your one half-empty bag of charcoal is spent.

The most important commodity will naturally be access to clean drinking water. Again, without stored water on hand in sufficient quantity, things will become truly desperate and fast! The human body can only go for a day or two without water before the organs and the mind start to fail. After a few days without water, our body starts to shut down. Death follows for the average person after a week for most without water. Inability to function normally comes before that with a 10% loss of body mass from dehydration considered a serious medical condition.[22]

So all we have to do is turn on the faucet in our homes or apartments, right? Wrong. After a large earthquake, the unprepared among us will try to fill up their bathtubs and any water jug they can find, knowing how important water will be. However, the fractured landscape that comes with major quakes often breaks water mains to entire communities. Water from a faucet may be unavailable within minutes after an M8.0 quake!

At that point, no one anywhere gets any water, including firefighters. Only those living on a body of clean freshwater (lake, pond, stream, or river, etc.) or have a well that can be pumped manually or via a solar backup power system will have reliable access to water. This assumes no one has taken possession of that water source by force, including the government. Depending on Mother Nature to deliver daily rain is a hopeless cause, especially in the western states like California where droughts and low rainfall are the norm. If you think you can wait with a bucket to catch the next rainfall, you may be dead from thirst long before it arrives.

Even with access to natural water sources, the need to purify water is key. How many of us have water-filtration systems to remove the harmful bacteria and protozoans that live in freshwater? Few if any of the people around you now, friends, neighbors, or coworkers, have any practical way of making sure the untreated water they may have access to is safe to drink. I recently canvassed a few online and big-box outdoor stores to see what would be immediately available for water filtration. Only online sources had any large family-sized water-filtration systems in stock. None of the several outdoors businesses had any at all. A few had the backpacker's portable survivalist-style individual water-filter units. Those who try to order anything online or get something at a store will be out of luck in the first minutes after a catastrophic quake. Store employees will buy the few individual units available anyway for their own survival. Again, like food storage, if you don't stock up before the quake, you won't be able to find what you need afterward.

In order to understand the kind of difficulties that you might experience after an M8.0 strikes, it might aid you to walk through just the kind of issues that are described above. How would you deal with long-term power outages, especially during winter? If you haven't already been through a large earthquake, volcanic eruption, hurricane, or flood, etc., where you have faced associated water, food, power, and security problems, doing this mental exercise would be helpful. Ask yourself how you and your neighbors might handle similar circumstances. Or will they depend on you for their safety and survival?

e. **Will Preppers be OK?**

The answer to whether preppers, including some of your friends and family who may already be into apocalyptic planning, will do the best as CGEs strike reveals some surprising answers. It isn't all it's cracked up to be.

Some preppers are very serious preppers. Some of these most dedicated preppers put the National Geographic TV series about preppers to shame. We will review in this section three classes of preppers to understand their thinking and see what we can learn. In the first category, we have fully prepared preppers (FPP) at the top of the list—those with significant resources and a comprehensive plan to survive the worst-case scenario. Then we have those middle-class preppers (MCP) associated with the middle class economically. People in this category have a lesser, though wide, range of resources and plans to make it through tough times. Finally, we have the short-term preppers (STP). These are the many who are thinking all they need is just enough preparation and a bag of essentials to get them through the first few days after a disaster until things blow over and they can safely return home.

Before we jump into these three prepper categories, it's important to discuss a vital subject:

Don't tell anyone you are a prepper! If the word is already out, don't discuss your specific planning!

The best way to have your neighbors (or the government) come begging (or appropriating) from your prepper stores at your home is to make it known that you are a prepper! If you are asked by a newspaper or TV reporter for your ideas on disaster preparation, do not get involved and thus let the world know what you are about. Once the rest of the people on the street are aware you have a year's worth of long-term food and water storage as well as backup power, you are in trouble. They will begin to think if the shit hits the fan (SHTF), all they have to do is go to their good friend down the street, and out of the generosity of his heart, he will care for their family. As a result, they will not prepare. Instead, *you* become their preparation. They will make the easy, lazy, irresponsible decision where they don't have to lift a finger or spend the first dollar.

So all of a sudden when the power goes out and the radio or TV says there will be no food or water for weeks or months, guess what happens? Your immediate and distant family members and good friends from three or four homes in either direction who know of your stash suddenly show up. Because you are a nice, sharing person, you give in. In a day or two at the most, your one-year food supply is cut to one or two months. In the following two days, the people two or three streets over have heard of your stores and your generosity, and so they also show up. If you are still a Good Samaritan, your one-year food supply is now down to a week.

Then things get serious. That side of town has heard you have "plenty" of food, water, and power to share freely. The desperate hordes by the hundreds show up and demand what you have. When you say you have given away all you had but a few days or a week for your own family, they don't believe you. Then they break in with force and weapons to steal what you have left. In two or three weeks, your one-year survival package is gone, and if you are lucky, you and your family will still be alive if your ransacked home hasn't been burnt down by an angry mob. Here is the most important message about prepping repeated:

Do not tell anyone you are a prepper, especially if you are an unarmed prepper and not willing to defend your home and preparation supplies.

Back to the categories of preppers.

(1). Fully Prepared Preppers (FPP)

In the class of fully prepared preppers (FPP), we find those with substantial resources to achieve near-utopian prepper existence regardless of CGEs, economic collapse, or whatever life throws at them. They have a wide range of capabilities to deal with whatever threat they are planning for. These are often wealthy people with remote mountain cabins or large isolated homesteads, family farms, or similar sanctuaries that are fully equipped and stocked with whatever is needed to live independently for a year or longer away from the masses and are often planned for more than one family. Most FPPs are highly educated, serious-minded people. They are not prone to conspiracies or "Planet X" type myths. Many have retired from successful companies if not their own companies.

Unfortunately, (or the others for that matter) are not planning for the kind of cold climate scenarios, including large earthquakes and volcanoes (excluding Yellowstone). Their greatest concern is more likely an economic collapse that would dwarf the 2008 crisis and recession or even the 1929 Wall Street crash and subsequent Great Depression. They may also be concerned with the real threats of foreign governments or disaffected individuals who can bring down the grid or major banks with a cyberattack. This book will, however, serve to reinforce their planning.

FPPs are characterized by the following:

(a).They are completely off the grid (or capable of quickly leaving the grid) and have their own adequate primary and backup power source(s). These FPPs typically have solar systems with overly large battery storage to withstand numerous long cloudy days in a row. These systems can easily run up to $20,000 to $30,000 and are often supplemented with backup generators with fuel tanks or wind systems which may be augmented by natural habitat heating and cooling built into their homes. What I mean by the latter is that they have specially designed homes that take advantage of direct solar heating during the winter months and avoidance of the Sun during the hot summer months. Some have resorted to earth-covered units or underground homes while most are still above ground. Others have gone to extremes in this category. They are the ones who go into abandoned missile silos usually in the far north latitudes where the Cold War missile silos are available for purchase. I don't favor this approach to becoming an FPP given the coming cold epoch.

(b). They have substantial food and water in storage. Most government scenarios involved in CGEs deal with the need for food and water for a few days to a week and are the kinds of scenarios the U.S. government, FEMA, and local emergency management agencies (EMAs) recommend. I believe these low levels of food and water storage are wholly inadequate to the nature and magnitude of the coming earthquake threats.

Word to the wise: *Do not follow U.S. government guidelines for catastrophic earthquakes—they are seriously inadequate as a rule and are based on the government and utility companies restoring services within a short period of time.*

Some of the FPP food storage units are impressive in size and styles of food being maintained and are usually at least a year or more. They can be expected to have extra stores in anticipation of direct family members who will come knocking at the door during the worst of times. In extended family FPPs, their planning actually is actually dependent on several families cooperating, including closest friends, their children, their families, and grandchildren. Multiple-family survival sanctuaries as a general plan are preferred. There are several reasons for this thinking:

If indeed the SHTF and you have to go into survival mode and defend your family and home with its food, water, and power supply, you will need others to help. Unless you have qualified adults to help through three shift operations, you will soon be exhausted, effectively manning a one-man guard post 24/7 for days at a time. Forget doing so for weeks—you won't last without backup. In addition, multiple families means more force to defend the sanctuary along with added skills and capabilities for long-term operations much less survival in a worst-case situation.

Water at the homes/sanctuaries of FPPs are in the form of stackable, somewhat-easily-moved specially made non-BPA containers. BPA is the unhealthy chemical that some plastic containers have that leaches into the products within the container over time. One should not forget that water is heavy! Anyone who has bought a standard wrapped package of 24 bottles of purified water from their local store knows that. Yet this store-bought amount of water lasts only for a short while. FPPs have sustained water sources where possible. These are permanent ponds, streams, and lakes that for most needs are limitless water sources. Deep 400-foot or more wells with manual and backup power for pumping are the most common means of access to water for FPPs since natural external water sources are in short supply and will be frequented by the many people seeking fresh surface water.

(c). They have the ability to defend their food, water, and power. Social order may break down in large towns and cities within days or weeks of food and water supplies being exhausted. Some preppers have gone to elaborate lengths to have their home fully stocked with food and water in a crowded subdivision without a security plan. When asked whether they would be willing to defend it by shooting their neighbors or criminals who are breaking in their door to kill then and steal their food, they say no. Further, some of these people don't like guns and don't plan to get any! These preppers are not the FPP type by any means.

There is no reason to be a prepper to any extent for an apocalyptic event unless one is mentally and physically prepared to defend one's life, home, and the means to feed and care for one's family.

As noted above, if your neighborhood knows you have a store of food and water, they will be beating down your door in short order. This is a certainty if they also know you are unwilling to defend it. Like the local grocery store, an undefended home will, by the pressure of compassion or friendship or force of arms, eventually give up its food and water if not their very home. These undefended homeowners or hapless would-be-preppers will be like everyone else at that point—victims.

FPPs, on the other hand, tend to be realistic apocalypse planners. They are, as a rule, heavily armed, well trained in firearms use and safety with ample ammunition stored. They will be the best prepared for self-defense, anticipating the eventual and highly probable lawlessness that will result during a long duration aftermath of a catastrophic event, including an M8.0 earthquake. Human nature has not changed for hundreds of thousands, if not millions, of years. The survival instinct takes full control during desperate times, and the predicted period of multiple catastrophic earthquakes starting as early as 2017 will create truly desperate times.

Those who fail to come to grips with this fundamental law of nature are seriously self-deluded and likely be among the first to perish during a sustained post-CGE period where there is no food, water, or power for weeks or more. At that point, survival becomes the number one mission regardless of politics, religion, or social mores.

(2).The Middle-Class Prepper (MCP)

This economic category of preppers is where we see the greatest variation among those who are preppers. It ranges from those who are practiced in family outdoors living via camping, hunting, fishing, environmental tourism, wildlife photography to simply taking periodic trips to a nearby park or camp ground—you name it.

Some of the best examples of those most-prepared MCPs are those with RVs and motor homes in the $25,000 to $200,000 or more price range who already have a piece of land of their own or shared with another friend or family, perhaps someone with a farm or woodlands. They are well educated in survival requirements and have ample resources and a ready sanctuary available where they can exist for months with their home on wheels with supporting food and water either brought in or, in the case of water, available off the land.

Right behind this group are those who regularly do extended camping outings with their families. This is one of the better ways of prepping for anything that requires evacuation from a major metropolitan area. Boy Scout organizations or other organizations that practice family joint activities in the woods, out on the prairie, or other environment where survival skills are taught are ideal. It is one of the best ways to acclimate a family, mentally and practically, for those times when the SHTF and one has to get out of Dodge in a hurry and then survive remotely away from city or urban life. Those who have made a practice of camping are already aware of the value of creating family experiences and parenting. These good reasons may come into play later when things get serious if one has already been exposed to the basics of living off the grid and relying on one's self for survival. Family groups like these have large tents and camping gear with established practiced means for long-term off-grid situations and have taught other family members that survival outside the big city is possible. Those who have not had this beneficial experience are not only in shock but also totally unprepared for outdoors living if they have to head for the hills for the first time after an M8.0 destroys most of the lifestyle they knew previously.

The other group in this category are those 13.7 million hunters and their families who are part of the 6% of Americans who are into rugged outdoor activities such as hunting and fishing, especially where such events are days or weeks long requiring multiple outdoor skills and resources.[23] Within this group, one finds those who have many experiences living out in the woods or on the river or the lakefront. They may have taken part in hunting and fishing events with friends and family or sometimes requiring professional outfitters who charge thousands of dollars for services associated with bringing home that big ten-point buck or that limit catch of brown river trout.

Oftentimes this group is associated with others in the hunting and fishing profession with whom they can join forces in difficult situations. Ask yourself. If you have to team up with another family, wouldn't you prefer it be with one that has the knowledge and equipment to fish and hunt wildlife for food and not only survive off the land but also as part of their hobby or profession, come fully equipped with firearm skills, weapons and ammunition to make it happen?

The last group in this category is Americans who are ever watchful and protective of themselves and/or their family and have developed a comprehensive plan for an emergency evacuation within their limited resources. These may have nothing more than a pickup truck or family van with another vehicle and/ or trailer in tow, yet they are practiced to a basic level in independent survival for weeks or months with the supplies, arms, and will to stay safe. This group is much more than the final category of preppers, the short-term preppers. This group may have the least of the resources of the MCPs, but they nonetheless are concerned and prepared to do what they can for their family's safety.

A challenge all MCPs face is much the same as the rest of us—can they safely get to their sanctuary before the roads become impassable following an M8.0 earthquake? For the fortunate few, the best option is not to try to make a precarious long drive to get there. Start living there now. That way when the rest of the family shows up, they will be welcomed to a sanctuary that is already at least partially operational.

(3).The Short-Term Prepper (STP)

In this much larger community of preppers, we have those who bought into the idea that when the SHTF, all they will need to do is grab their bug-out bag and head for the hills for a few days until things blow over. There are some fundamental problems with this approach to how one can survive the next CGE, especially if an M8.0 has laid waste to the water, power, and transportation infrastructure of an entire state or several states of the USA. Here are some of the flaws I have found for the STP and why I do not recommend this as one's only survival plan:

(a). Lack of supplies: The typical three-day supplies in the bug-out-bag will quickly turn into zero supplies even if one can find a safe place to stay for a while. At this point, the STP is back to square one and quickly becomes another member of the rest of the hungry, thirsty mob. Just three days does little in the face of weeks or months of deprivation unless the three days are a bridge that permits one to get to a larger safer location with long-term food and water supplies.

(b). The difficulty in bugging out: Should a CGE in the form of an M7.0 to M8.0 or even an M9.0 earthquake take out much of the infrastructure of one or more states, there will be a mass exodus from large metropolitan areas. Most will figure out in short order that the larger the city one lives in, the larger the number of people there will be resorting to pure survival instincts—just the kind of situation and people one will need to get away from. Large inner cities with a large percentage of people totally dependent on the government for food and water who will suddenly have none of either will quickly become hundreds of thousands if not millions who will be capable of doing whatever they must to survive. This resultant huge rapid evacuation of cities will quickly clog all highways and interstates full of desperate, unprepared people leaving for parts known or unknown. It will make rush hour look tame by comparison. Major highways will quickly become parking lots with no one going anywhere. This situation is all but guaranteed by the current administration's lack of planning and lack of education of the people in these high-risk zones.

What that means in a practical sense is that the STP will be stuck for hours and/or days on the roads like so many other STPs or various scared people. Some may be able to drive around jammed highways with the right kind of off-road vehicle, but most will not. Most of those with whom they will be stranded on highways will soon lose hope. Some may have hastily thrown a few items of food and water, a pistol, or rifle, etc., into their cars, trucks, SUVs and sedans, or motor homes. But because they did not plan on an evacuation properly, much less practice such, their bug-out bags will probably be a hodgepodge of supplies that only a few minutes of unplanned packing could produce. These people and their STP partners will also soon be out of everything they will need for long-term survival and will again become victims.

(c). How will you survive at your bug-out location? Is any place safe? So if you are an STP and you head for the hills, what will you see when you get there if you can by some miracle? Chances are that the hills or mountains or wilderness you have identified as your temporary sanctuary are already occupied before you get there by the dozens or hundreds or thousands who have the same short-term plan and a BOB just like you. This assumes the property owners give you permission to set up a tent or park your pickup, family van, or mobile home there.

And after the multitude of STP survivors have finished off their BOB food and water supplies, then what? If you are fortunate and have plenty of water via a pond, lake, or stream and a water filtration system, that's a big step ahead in survival. But what about food? You brought your 22LR rifle and 100 rounds of ammo and can feed the family off squirrels, rabbits, and birds, right? Wrong.

It seems everyone else has brought their small-game hunting rifle or perhaps something much more lethal like a shotgun or semiautomatic AR-15. Then in a flurry of small-arms fire over a few days, all the local game animals, if not some of the populace, have been eliminated. Now where do you get food? If you are into big game, a similar story—all the deer will be quickly shot and consumed or chased away in short order. Fishing, same story also. Available fishing holes will be fished out within a week or two.

Only the very fortunate with access to relatively unlimited offshore fishing or large ranches and tracks of land especially with herds of cattle or ample deer, elk or hogs, or other edible livestock that are well defended can make this STP survival story work for them and their family. There are very few of these locales with ample food and water and certainly nowhere near the amount needed to support the millions who will need such a well-supplied sanctuary.

It's important to understand a fundamental rule in the live-off-the-land or off-the-grid scenarios. The vast majority of Americans get their food at the grocery store, not out in the woods or off the farm. Moving down aisles in stores is their only proven skill level when it comes to obtaining life-sustaining food, and not slaughtering a pig out of the pig pen behind the barn, much less moving down ravines and streams or over mountains stalking wild game. Forget raising vegetables after the quake strikes—you'll be too late for that approach. To expect a large segment of the population to suddenly seamlessly, efficiently transpose their lives, even for a few weeks, to operating safely in the woods, farms, ranchlands, or campgrounds and do so successfully is unrealistic. For many, it could be fatal.

With no survival skills and few resource options, the great majority of those near the epicenter of the catastrophic regional earthquakes predicted in this book will revert to basic life-sustaining measures in the first hours and days after the quakes strike. It may get ugly and brutal, and the suffering will last for a long time. The best chance an STP has is to get out of the affected area and, if possible, make it to another state that is unaffected. That should be the strategy of the STP. The BOB should be viewed only as a short-term resource to enable the STP and the rest of the family to get to where it is still relatively safe and food, water, and power are available.

(d). How long can you survive? What happens if things don't just blow over in a few days?

After an STP arrives at a place in the woods, the desert, the mountains, or a nearby farm or ranch, how will he/she make it as days run into weeks which run into months? When winter hits with a vengeance, if it hasn't already, how will the STP and family get through these dangerous times. These are basic questions that one must confront before believing that being an STP is a solution to the long-lasting problems millions of Americans will face as early as tomorrow, next month, or next year after the first M8.0 strikes.

If you are an STP, please accept this section of the book at face value. If you are an STP with just a bug-out bag and maybe a six-pack of water bottles and haven't planned far beyond the first 72 hours, you may be in for real trouble. In just a few days, you may find yourself stuck in the boondocks with everyone else hungry, thirsty, and truly desperate with no remaining options.

(e.) Flying out of danger: One can forget the large airports with communications systems, power, and fuel lines down from a devastating regional earthquake. Even with airports with manual emergency takeoff procedures (few if any other than small airfields), that assumes that the runway is serviceable. Support personnel and flight crews may not be able to get to the airport if one is still operating, which will not be the case close to the epicenter of a catastrophic earthquake.

Has anyone at the Federal Aviation Administration (FAA) certified all West Coast or NMSZ or Charleston control towers standing high above the airport to withstand a range of M7.8 to M8.0 or larger? I'm afraid the answer may not be a good one. Helicopters may be operating but only for short durations until they can find new fuel locations that are secure. But how many of us have a chopper standing by, and how many have a pilot and copilot lucky enough to make it to the airfield with chaos on the roadways without their own family in tow? The same issues apply to all the private fixed-wing pilots, their aircraft, and passengers. Can they all get to their planes, and will they be able to get off the ground if they do? Similarly, with no control tower and/or a broken-up runway, no one will be taking off or landing.

(f.) Complete lawlessness within a few days: The loss of power, food, and water with no quick reassurance to give people hope, along with major utility and critical infrastructure damage after an M8.0 strikes, would lead to the rapid erosion of law and order. The situation could be so bad as to give many individuals no alternate but to discard long-established civil and social norms to stay alive. The few police and special security personnel that are normally available for street riots and unruly protests would be no match for the widespread general lawlessness that comes with a complete breakdown of all economic and social structures that may persist in a catastrophic earthquake aftermath that is months long.

(g.) CGEs also have the bad habit of creating collateral and cascading disasters: Earthquakes beget tsunamis and violent ground shaking simultaneously which in turn begets more earthquakes via powerful aftershocks. This means if you are in downtown Seattle if the full Cascadia Subduction Zone (CSZ) breaks bad, and if you are not crushed by a falling building, you are likely to drown fifteen minutes later in the tsunamis that destroys what is left of lower Seattle. If again you are blessed and thus survive these horrific initial events, the chances are high that you will not survive much longer without clean water and food, which all remaining people in the greater Seattle-Tacoma region will then be scrambling for, fighting for, and dying for. Then for the remaining few, polluted water supplies, fires, and widespread crime and gang activity will be omnipresent. History suggests that disease outbreaks accompany contaminated water supplies. If you do not plan for this level of societal breakdown and resource shortages, you may come to regret it. If you do and no quake happens, then your life and that of your family will be more secure for the many other tragedies that may strike.

During periods of increased illness from the direct and indirect consequences of an M8.0 earthquake, how will hospitals cope with the historic numbers arriving at their emergency rooms? Without power, they will be unable to conduct lifesaving operations. FEMA studies noted in previous chapters indicate that many hospitals will also be damaged or destroyed in such a catastrophic-earthquake scenario. With power outages and transportation and communications systems down for extended periods of time, how will banking and other electronics-dependent systems keep operating? Of the major players in this financial area, how many have actually tested their primary backup systems to withstand an M8.0 quake? There will be a myriad of indirect effects from a catastrophic statewide or multistate earthquake that will eliminate what our modern life now provides. Where will chronically ill people get the life-sustaining drugs they need when drug stores have been looted and have no power after a regional M8.0?

h. Most Americans are unable to prepare even if they are aware of the threat: Recent data on the financial health of Americans is appalling. This says for many, there is nothing they can do even if they accept everything we say in this book.

As of June 2016, the government's Supplemental Nutrition Assistance program (SNAP) had 43,376,981 Americans in just over 21 million households who subsist on food stamps.[24] Though that number is down from 47,636,090 in 2012, it's still a huge number. This truly ugly stat says a great many of our citizens (about one out of seven) are totally dependent on the federal government and the electronic systems that control their access to food. When the next cold climate begins to take a toll on crops and/or if one's region is devastated by a major earthquake where the grid goes down and stays down for days and weeks with no computers or cash registers working, how will these people get food to survive? Further, to expect these same poor people to make any preparations that will see them through a predicted period of months without any assistance is not based on reality or human nature.

Financially, many Americans simply do not have money today to devote to food or other preparations of any kind. According to one study in 2016, the typical family in the United States cannot come up with as little as $1,000.[25] I have looked into this issue. This lack of funds for millions compares with any of the small long-term survival food storage companies that charge a couple thousand dollars for food for a family of four for 3–6 months. And most of these expensive long-shelf-life food packages are high-carb diets with the only protein supplied by beans. Meat protein packages are much more expensive. Again, this segment of the population just does not have the ability to prepare for what is coming.

Then we have those who cannot find work or have decided to stop trying to support themselves. The latest figures from the U.S. Department of Labor for August 2016 report the labor force participation rate was tragically high at 62.8%. This translates to 94,706,000 Americans have left the workforce altogether![26] The Labor Department defines these masses of unemployed as "persons who want a job, have searched for work during the prior 12 months, and were available to take a job during the reference week, but had not looked for work in the past 4 weeks."

These unfortunate souls live off family and friends, part-time work, and government welfare or unemployment benefits.

In other words, approximately 95 million of our citizens (about one in three) are living off friends, family, their community, or the government, i.e., other taxpayers! This is the shocking number of Americans who will be the first to be impacted by the devastating economic impact of regional earthquakes predicted in this book. Almost overnight, millions of these people will have their lifelines cut off. They will be without resources for basic needs. How will they react?

How will they deal with the increased employment despair that will come from a regional economic collapse during a CGE? If the friends and family they depend on now are suddenly out of work and without a paycheck because their employers' business was destroyed or closed or because supply lines for it to operate are shutdown, how will they make it? How will they receive any government payments if the banking system and food payment system in their entire state is shut down for weeks or months? This situation compounds the already-deplorable employment/ dependency story we have across the USA.

How Will Businesses Deal with a Catastrophic Earthquake?

This is actually a more complex issue because it deals with the technological complexity of our society which is both far ranging and in-depth in our day-to-day lives. Certainly we must face the prospect that the majority of business within the proximity around the epicenter of an M8.0 plus earthquake will be either totally destroyed or out of business for an extended period of time. This means no income and no jobs available for employees; therefore, no way for these workers to provide for their families likely for months to more than a year. High-tech companies with precision high-dollar equipment would see the longest recovery times since many machines and specialized tools for their industry take one to two years to fabricate and often must be shipped in from across the USA or from other countries.

Some companies in the medical and high-tech area are one-of-a-kind businesses with no effective replacement in the marketplace. Drug companies developing the next generation of drugs and new technology start-ups could lose years of work and hundreds of millions of dollars in investment in the space of a 3–4-minute earthquake comparable to an M7.0 to M8.0 earthquake. Insurance companies with large client bases in the known high-risk seismic zones could see their worst losses in their history. Some insurers, other than multinational ones, may not survive. Despite the obvious threats the research from the IEVPC has demonstrated, our attempts to get the largest insurance companies in the United States to take this matter seriously have met with polite rejection.

It is astounding that even with the threat that exists, some insurers have their headquarters or major company data centers smack in the middle of these high-risk zones. This constitutes not only bad corporate governance but also contradicts their very mission. How can they properly assess the financial risks of the predicted largest threats to their clients and their own company if they don't acknowledge that the CGE threat exists right under their corporate headquarters in the first place? If a major business's insurer just down the street is not preparing for an M8.0, why should they? Isn't it the job of the insurer to understand the threats to their clients and to communicate those threats to them? Why haven't the largest insurers already increased their premiums to cover these risks? When these insurers claim bankruptcy after the first one of two catastrophic earthquakes put them out of business, who will come to their rescue and that of their policy holders? When business insurers cannot rebuild the assets of heavily damaged businesses, who will?

What about the stock market and other economic impacts? Depending on the location and extent of damage, any of several likely earthquakes identified by the IEVPC could result in a major impact on the U.S. gross domestic product (GDP). Further, the state of California has been battling budget shortfalls and huge pension and social costs for many years. Financially, the state is already out on a limb in terms of fiscal resources. A devastating M7.8 or larger quake anywhere along the San Andreas could be a coup de grâce for the state, causing the largest state bankruptcy in U.S. history. Similarly, northeastern states facing critical oil and gas shortages after the NMSZ erupts would also see substantial economic losses and long-term declines in tax revenue which would further add to the financial instability of those states with slim to no cash reserves and heavy obligations. In the face of the regional economic devastation any of the high-risk quake zones pose, how can the business community survive, especially when they have done no prior preparation for such widespread destruction?

How will the major airlines function when flight and airport communications go down and at best survive only for a few days before backup systems fail? Then they have the issue of infrastructure to deal with. Planes need smooth straight firm runways to land on. Earthquakes could make runways unsafe to use altogether, and their repair would take weeks or months or longer. Some infrastructure damage could prevent heavy equipment availability or delivery to do the rebuilding.

How will businesses be able to operate if many of their employees have not come in for work as they care for their homes and families and cannot or are unwilling to come in? If they went to work, would gangs of marauders break into their homes to steal food, weapons and ammunition, or anything they can use to barter with during weeks, months, or longer of difficult, often violent existence in a worst-case scenario?

Such scenarios are utterly impossible for most to comprehend since we have no collective or cultural awareness of destruction on such a scale. Of course, our elected leaders for the most part will do whatever they can to avoid confronting such real problems.

How Will the Department of Defense Deal with a Catastrophic Earthquake?

What about military and defense issues? Large-scale regional earthquakes could produce ground accelerations that could destroy military facilities, and in some locations, tsunamis could heavily damage if not wipe out port cities including their nearby U.S. naval facilities. Runways could be destroyed on air force bases not to mention loss of power for National Guard, coast guard, and army stations and posts that would normally be among the first called upon to mount rescue operations. This is a key point. In past emergency declarations, the National Guard or other military units have responded. What does the earthquake survivor do when these forces cannot show up?

The entire concept of going in after a catastrophic earthquake and immediately rebuilding may be out the window should the Cascadia Subduction Zone, the San Andreas, or the NMSZ deliver the magnitude of destruction they are capable of and are predicted in this book, along with foreshocks or numerous powerful aftershocks. For example, the NMSZ, during the winter of 1811 and 1812 had three or four catastrophic quakes in a row over a three-month period. If anyone had gone in to rebuild, what they rebuilt would be demolished by the next major earthquake in the series of catastrophic quakes. Could this scenario repeat between 2017 and 2038? Absolutely it could and with a probability of 80% according to our research. Local governments and national military units should be planning for just such an occurrence!

Military airfields, ports, and bases within the high-risk zones discussed in this book require pre-event exercises to determine whether military readiness is compromised by a single or series of catastrophic earthquakes. This is especially of concern for National Guard units within each state that would be among the first to respond to an M8.0 earthquake. Would they be able to?

Large earthquakes are not singular events that are over after the first ground shaking is done. It would not at all be unusual for an M8.0 (San Andreas) or an M9.0 (Cascadia) quake to be preceded by foreshocks or followed by aftershocks of M6.0 to M 7.0, which are quite powerful in their own right. The 2003 Haiti quake that killed at least 100,000 was M7.3. As we have already seen, the NMSZ is capable of delivering a series of catastrophic earthquakes of historic magnitude and destructive capacity. This scenario of a single M8.0 earthquake by itself is capable of knocking out power grids, bringing down large buildings, and breaking up interstates and main highways within large metropolitan areas. This is especially the case with aftershocks which would hurl their energy at already-weakened buildings and roadways.

These quakes do not discriminate. All large state and federal government agencies and their facilities could be heavily damaged and put out of commission after an M8.0 event. Many Americans are completely dependent on the U.S. government for their livelihood. Can the government agencies continue to deliver services on time after the predicted quakes in this book? It is likely they have never even run the first realistic drill under this scenario. Again, this means that not only should we not expect government assistance after an M8.0 quake, but we should also expect that they will be facing calamitous losses themselves for those agencies and offices within the high-threat areas! In other words, the local and state government offices that many frightened people might call on for help after an historic M8.0 earthquake may simply no longer exist.

Major underground oil and gas lines transporting gas across the nation could be torn apart, dumping their volatile products and resulting in numerous massive fires in affected areas.

How can repair crews and firefighters get in to fix them if the highways are in shambles? The fire threat in large cities could be hard to estimate but would clearly be a big problem for emergency personnel. Widespread broken gas lines and people making home fires to stay warm in the middle of winter would add to the numerous types and sizes of problems that governments and first responders would have to deal with on a scale that would be the largest since our country was founded.

Availability of hospital and other crucial medical services may be put out of action indefinitely. Seriously injured personnel may be unable to receive any major medical care or surgery if the local hospitals and clinics are among the buildings brought down or otherwise declared unsafe to enter. The injured will either go untreated or treated by friends, family, and neighbors with questionable medical skills. EMTs normally stationed beside firefighters would be tested like never before, unable to make even the smallest dent in the thousands or tens of thousands of calls for help they may receive in the first day after a CGE hits, compared to none or a few per day during normal work shifts. And this assumes some phone systems are operating so that 911 dispatch centers could have the first responders get there in the first place.

Unlike isolated quakes that Southern California has seen in the past, there would be no rapid assistance from neighboring counties since they may also be knocked out of commission by the same M8.0 or greater quake. In a regional earthquake, assistance may be one or two states away, not the next city or county next door. That means there could be long periods of time before the injured would be receiving the most essential and timely lifesaving aid. Imagine a statewide or multistate-wide Katrina with no one to provide aid for weeks or months, and you start to get the idea!

These are just examples of the truly challenging times that lie ahead for Americans. After all, as a nation, we have not had to contend with the scale of tragedy we are likely to face between 2017 and 2038. The initial reaction to some will be shock, disorientation, and a lack of acceptance of just how bad off they are. Sadly, many who have been used to the government paying for their food and otherwise being cared for by someone else will probably suffer the most.

Recommended Preparations for Individuals, Businesses, and the Government

The matter of how to prepare for these predicted CGEs, regardless whether they are in the NMSZ, South Carolina, the West Coast States, and Alaska, depends greatly on one's personal circumstance and thus does not lend itself to having a book such as this propose the myriad of preparations that fit each person. However, some basic recommendations can be made for generalized situations. By following the steps below, one can get a pretty good idea of practical survival methods that may be helpful if not lifesaving.

What Individuals and Families Should Do to Prepare

Here are our recommendations for preparations a family should take if they are now living in a primary threat area for a CGE or in this scenario an M8.0 earthquake. These guidelines are designed for a family of four defined as a working husband, working wife, and elementary school daughter and son with one medium-size dog, living in a suburb surrounded by hundreds of other homes, condominiums, or apartments.

Step 1. You should move your family out of the known high-risk zones as soon as possible!

There are no employment or financial gains or family relationships or community connections or allegiances worth the risk, like the lengthy suffering and possible loss of life that will come to you and your family if you are caught in a post-CGE environment. This period of life-altering distress will cause you and your loved ones to face a months-long or years-long ordeal of having to survive under the most difficult conditions imaginable. If you decide to stay where you are and ignore the warnings and recommendations in this book, you and your loved ones may be forced to endure unending hardship going on for weeks or months without adequate water, food, power, medical assistance, or basic security. You can always return a year or two after the CGE has done its damage and normal life reaches a reasonable level, maybe. If you relocate, who knows, you may even find a better, safer permanent environment for your family and not have to try to return to your old homestead or what's left of it.

If you do not follow step 1 then at least give the following steps serious consideration.

Step 2.Become educated in how earthquakes unfold and how to prepare either by going online or purchasing books or consulting local emergency management agencies (EMAs). Educate your family and encourage your neighbors to do so as well.

Step 3. Develop a plan to stay in place and survive the worst the CGE can deliver as long as you are in a relatively safe place.

That safe place is defined as not in a high-rise homestead, not living on a coastline likely to be wiped out by a tsunami of 30 or more feet in height. If you decide to stay in a high-risk CGE area and heavily damaged major metropolitan area, at least move to a safer location from which your family's survival may be maximized.

Step 4. Obtain and properly store a minimum one-year food, limited water supply, and medical supplies in your home.

Do not tell your neighbors you are doing so! Plan for your government-dependent family members and neighbors to call on you for help when you calculate your total extended family's one-year food and water requirements.

a. Home Planning. Freeze-dried foods with a 25-year shelf life are recommended. If you do not have funds to purchase a one-year food supply, start small and build up your food reserves. Set aside an area of your home or apartment as an emergency storage area for food, water, batteries, etc. You may be surprised how quickly at least three months' food can be assembled beginning with a few cans a week and cheap long-lasting bags of rice. When there are no food stores open after a catastrophic quake, what little you have set aside beforehand may be lifesaving.

   The amount of water supply on hand is a function of usage rates and the possibility of replenishing it without difficulty. Store water in easily stored/stacked sturdy, non-BPE water containers that can be quickly moved to your car or truck if you have to relocate. Local water sources may be quickly contaminated. Water utility companies may be unable to deliver water to homes and hospitals for weeks or months because of power loss and fractured water mains in numerous locations.

   Portable water purification systems are recommended. Do not wait to purchase these critical items. Get your water-filtration systems now, today! Ensure you have adequate medical supplies for emergencies and normal medical treatments. Get drugs for those already under treatment for a period of six months to a year if possible. If funds are limited, immediately start a plan to slowly build up food and water and medical supplies and emergency equipment until you reach your one-year supply goal. Even if you only have three months' food and water before the next catastrophic earthquake, that's three months more of life and possible options you would have as opposed to having nothing to survive on and no options.

b. At-Work Planning . This step includes maintaining a three-day food, water, and essential supplies storage or bug-out bag (BOB) in your car or truck at all times between 2017 and 2038 if you live in any high-risk CGE region. We should assume that the vast majority of employed people will be caught at their work locations or in transit between work and home during the predicted CGEs and may not be able to get to their home or other secure site for many hours or days given complete loss of power and heavily damaged local transportations routes immediately after a CGE strikes. Cars without off-road driving capability may be caught in effectively permanent traffic jams of major highways once highways are broken up and made unusable by the expected M8.0 earthquake. This BOB may provide survival capability to get home or to another safe location on foot. Update your three-day survival kit as the seasonal conditions vary, i.e., hot versus cold weather.

Step 5. Prepare to Defend Your Home with Lethal Force.

Obtain small arms for the protection of your home, family, power, food, and water supply. Become practiced—you and your spouse and, if appropriate, older children—in the safe use of firearms. If you have a long drive to and from work, obtain a pistol with concealed-carry permit for use with your BOB in accordance with state and federal laws.

Step 6. Obtain backup power capability.

Have backup power at least in the form of generator with gas cans and/or rooftop solar power or wind systems. Ensure that adequate supplies of lights, batteries, and other essential supplies are available for an extended period of weeks to months. Batteries and power storage units that can be recharged from small portable solar power systems may be lifesavers.

Step 7. Be prepared for substantial social and community turmoil.

Be mentally prepared for instances where friends and neighbors resort to anti-social if not criminal behavior. Prepare for the mental shock your family will experience following the geophysical shock. Discuss this openly with family members to minimize the mental trauma that may come from them observing hostile behavior and a breakdown of social order. Remember, there may be no law enforcement, fire department, or military to come to your aid in the event of regionally destructive earthquake at least for the first few weeks. It will be best to assume you will receive no aid and plan accordingly.

Step 8. Develop and practice an evacuation plan.

Leaving a metropolitan area may become a necessity after a CGE for reasons discussed above. This plan must be quickly implemented should all your preparations not provide sufficient safety for your family at your current location. Practice your evacuation plan via several routes to a distant sanctuary. Ensure the owner of the sanctuary is fully on board and supportive. Aspects of this plan require other practical steps like always having at least a half tank of gas in your vehicle and spare gas, safely stored, if possible. With roadways destroyed or backed up with fleeing survivors and chaos in the streets, it may take a day or two to make what used to be a thirty-minute interstate drive to your home from work before you actually begin your evacuation. Make your plan accordingly, and if the worst-case scenario does not unfold, you will still be better prepared to deal with whatever happens..

What Businesses Should Do to Prepare

For business owners, here are some practical recommendations regardless whether your business is large or small:

Step 1. Relocate your business at once—away from the high-risk CGE zone you are living in.

Do not delay. This move will not only guarantee, in most cases, the survival of your company, but it may also provide the essential resources and safety your business will need and deserve. For publicly traded companies, remember, your stockholders require you to assess all likely risks and take actions to protect if not improve the value of their investment with or without a regional earthquake.

If you decide to keep your business where it is because of customers etc., then the following recommendations may assist you and your business to survive the aftermath of a CGE.

Step 2. Develop a plan to operate your company in the event of a partial or complete loss of operational capability.

This entails the creation of an additional component of your company to maintain operations at a contingency site. Specifically, establish a distant facility from where company operations can continue should your primary facilities become out of action after the CGE strikes. Ensure all required databases and communications for your company can be operated from the contingency site and adequate supplies, communications, and transportation networks exist from the contingency site. The use of the contingency site should be thoroughly exercised so that transfer of company operations from your existing offices can be made to the contingency site in as seamless a manner as possible under the expected extreme circumstance associated with an M8.0 earthquake.

Step 3. Ensure irreplaceable data and technology and documentation are copied or easily replaced.

Ensure that an extra set of important proprietary information and technology is available at the contingency site so that the company can be reconstituted at the original home base after the worst effects of the CGE have passed and rebuilding can begin.

Step 4. Ensure all employees are part and parcel of the contingency site planning and their own individual CGE preparations can support your company.

Conduct in-house training and education programs so that a maximum number of your employees are on board with the program and will actively participate in keeping the company operating. To the maximum extent practicable, make them and their family part of this plan and its exercises.

Step 5. Install backup power and fuel for when the grid goes down.

This will allow you to keep operating until restoration of permanent grid power. It will be important to verify that this backup power is installed such that it is also not taken down by the CGE. Remember the Fukushima quake/tsunami lesson. It was not the initial quake that caused the reactor meltdowns—it was the loss of the backup generators from the tsunami that caused the three of the four nuclear reactors to explode. The generators, damaged by the tsunami, were unable to keep coolant flowing through the reactors after main power was offline from the quake.[27]

Step 6. Meet with your suppliers and cooperating firms to bring them into your planning to survive the predicted CGE.

Let them know how you intend to keep operating and keep your relationships with them going. This will force the wise among them to join with you as they develop their own business survival plans.

Step 7. Meet with or somehow inform your customers of your plans to continue to be there for them after a CGE strikes.

This will serve to ease them into the issue and will show how well your company is being run that you will still be there for them when and if a CGE strikes. This may enhance your sales and credibility among those who are open to the idea of being prepared. This is particularly important if your company provides services or products that will be most needed after an M8.0 quake strikes.

Step 8. Develop plans to have adequate supplies on hand to get through the first weeks and months when supply chains will be broken and/or only partially functioning.

Step 9. Conduct dry runs where and when practicable to check on business capabilities during total (or reduced) operation stoppage from the primary site as well as ensure that the backup power will function when needed.

Step 10. Develop a security plan and test it.

This plan is similar to those needed by individual homeowners and is in anticipation of widespread lawlessness that may result from prolonged lack of food, water, and power following a CGE. Once word gets out that you have fuel and power and are still operating, someone may come to take your resources away from you!

Step 11. Develop new income streams.

Where possible, recognize that even during a natural disaster, there are ways to enhance business income. Examine how that can be done for your particular business and at the same time aid your customers. If possible, add new business operations oriented to recovery activities within your company's existing business plan. Make sure they can be quickly scaled up with reliable supply lines after a disaster occurs. Remember, hundreds of billions of dollars in reconstruction and emergency services funding will go into the damaged regions. Will your business be in position to take advantage of this windfall income?

Businesses that begin ASAP to prepare for the coming CGEs will automatically get their employees and thus other individuals and families to prepare as well, making the community more resilient following a CGE. Businesses that begin ASAP to plan to survive if not thrive during the coming CGE risk period will be those who succeed and rise above their competition.

What the Government Should Do to Prepare

For the past few years, there have been numerous internet reports of the Department of Homeland Security (DHS) and FEMA buying up long-shelf-life food and unusual amounts of ammunition far beyond normal requirements and other reports of FEMA camps for internment of large numbers of people. While much of this conspiracy-theorist scare mongering cannot be verified, there are plenty who believe there is evidence of a government plan to be prepared for some unspecified cataclysm while making sure the people are kept ignorant.

It is easy to see why so many do not trust the federal government and have fallen into the conspiracy theory mind-set. From a climate-science and earthquake-prediction perspective, it is easy to accept that the U.S. government cannot be trusted. The United Nations has been deceiving a full generation of us for at least thirty years on the matter of causes of climate change. There is ample proof of that.

Here are our general recommendations to the U.S. government to prepare for the multiple catastrophic geophysical events we are about to confront.

> *As a minimum, a massive nationwide restructuring of the existing responsibilities within our government to prepare for the predicted natural catastrophes should begin. This should be similar in scope and nationwide dedication and seriousness of the planning that was done during the Cold War years when the United States was expected to come under nuclear missile attack from the former Soviet Union.*

Regardless of what is going on in the inner circles of the White House, the Department of Homeland Security, and the Department of the Interior, here are some basic steps our next president and the rest of the government can take to get us ready for the catastrophic geophysical events that are predicted in this book:

1. Start by telling the truth about the real causes of climate change so that our people no longer live their lives and prepare for the future based on politically driven bad science of man-made climate change. End all climate-related work and funding dependent on the greenhouse gas theory in favor of solar and solar system-driven climate science—the most reliable available.

2. End the archaic and destructive policy within the USA that no one can predict earthquakes. Immediately begin programs to improve earthquake prediction, drawing on all international seismology experts.

3. Assemble a geophysical emergency council (GEC) or similar entity within the executive branch to have the sole mission of preparing the country for the coming CGEs. Each agency and department of the federal government, including the Department of Defense, DHS/FEMA, and the DOI/USGS, should be represented.

4. Give a national press conference from the president and announce how serious the threat is and some of the key steps the federal government is taking to get our people ready.

5. Announce that every citizen in the most earthquake-prone zones will have to be specifically educated and prepared to deal with the predicted threats for their region.

6. Develop specific guidelines for each threat zone (a survival handbook) for each resident.

7. Develop an independent outside inspection and feedback mechanism for evaluating the nationwide readiness to handling the predicted CGEs.

8. Plan for and then conduct exercises on a massive scale to validate timely responses to the occurrence of predicted CGEs for each threat zone. This should include federal and state plans of threat states and their adjacent states to render aid once the CGEs strike. This planning should assume a worst-case scenario will unfold. Accordingly it should be concluded that each state's resources will be either totally compromised by

the likely CGE or will have quickly exhausted its capability to deal with the aftermath of the specific CGE.

9. Collateral damage planning and drills should be taken in areas that may indirectly affected by the predicted CGE.

10. The U.S. Department of Defense will have to be an integral part of the planning and be a major supplier of security forces and logistical support following a CGE. The DOD will itself have to deal with the destructive impacts of these CGE on regional military installations. This applies specifically to U.S. Air Force and U.S. Navy and Coast Guard bases where they may also sustain heavy damage, thus limiting their military readiness and their ability to assist in the recovery in states hit by these CGEs.

11. The federal government and the GEC should also assume that, along with the predicted CGEs, the time table for these events will coincide with a historic decline in global food supply. This is the result of a dramatic decline in solar energy, a solar hibernation, or grand minimum. Solar cycle 25 which begins around the year 2020 marks the full-fledged solar hibernation cold phase of the 206-year solar cycle. Research suggests that both cycle 25 and cycle 26 will define a 30-year long cold period. During this historic time period, the human race will have to adjust to far lower crop yields than the bumper crops we have seen up through 2016. The progressive transition from the past Sun-caused global warming is already showing new record cold and snow events along with the fact that there has been no effective growth in global temperatures for 18 years plus.[28]

12. The president and the GEC should recognize that the deployment of large numbers of security personnel, food, shelter, and water and portable power systems will be required immediately after the first CGE strikes. These recovery operations may be the largest since comparable scale operations during World War II or the Berlin Airlift. Preparations for these national recovery operations should be conducted for each of the CGE high-threat areas in the U.S., and they should be done, especially in preparation of both the winter and summer months when weather conditions will be most difficult.

13. Establish as the primary means for providing water and food for affected regions, a home resource program. This program requires people to establish their own post-earthquake

stores of food, and water, and where possible, power for at least six months' duration. The government should not be reviewed as anything other than a slow deliverer of limited supplies after home and local resources are expended.

The most effective means for assuring people in high-risk quake zones will survive the immediate aftermath of anything approaching what we have predicted is decentralization of resources down to the household level, and not centralization at a federal level. The government cannot replace the capability and reaction time of every household among the millions that will be affected, where those households already have their own emergency food and water supply immediately available after a catastrophic earthquake. It would take weeks at best and likely months for the U.S. government to mount such a massive regional resupply campaign during which time many citizens could perish.

FEMA put out a handy guide in 2004 for short-term food and water stockpiling that represents a start, and only a start at the kinds of planning individuals and families should be doing to prepare for a quake. This guide is available online for download[29] and is intended for short-duration natural disasters like earthquakes, hurricanes, and winter storms. It does not consider the scale of devastation or the length of recovery mentioned in their study of the NMSZ, much less our predictions. Since it is a government guide, it has several statements you should discount. For example, the guide leads off with "Even though it is unlikely that an emergency would cut off your food supply for two weeks, consider maintaining a supply that will last that long." There are many other preparation guides online for those who want in-depth coverage. However, each of them must be reexamined in light of the recommendations we are making in this book.

Just imagine the millions of families who are awakened in the middle of the night to their homes falling down around them only to find out that the water systems are destroyed and they have little or no food or water until a government response program might provide assistance that at the least will take weeks or months. The only way to care for these millions without food and water is to encourage, if not require, some of them to maintain an at-home or at-work, if not in-the-car, food and water supply to get them through the initial difficult days and weeks. The last thing we need is another bloated, unresponsive, wasteful federal or state government agency trying to store enough food and water to last tens of millions of people for a period of months.

14. The Department of Energy (DOE) should conduct an immediate risk assessment of all nuclear reactor facilities within the states at high risk of catastrophic earthquakes, tsunamis, and volcanic eruptions. They should verify that design specifications for as-built nuclear facilities meet the highest expected magnitude of the predicted CGE for their region. This assessment should include all reactors and their backup power sources to preclude a Fukushima-type disaster. The assessment should also include storage of spent fuels rods and other nuclear waste storage facilities. The DOE should publish a public status report for the public's review. Nuclear reactor facilities that cannot withstand the upper magnitude of the predicted CGE for their region should be safely decommissioned on an emergency procedure basis and all nuclear fuels and waste relocated out of these high-risk regions ASAP.

15. The DOE should ensure that all gas and oil pipelines that transit high-risk regions are equipped with real-time monitoring and emergency shutdown capability sufficient for the maximum predicted CGE threat. This shutdown capability should be independently evaluated.

16. U.S. states in the northeast in conjunction with the federal government should immediately develop a means for coping with a complete loss of the oil and gas normally delivered through NMSZ oil and gas pipelines so as to minimize the direct and indirect damage and economic losses and loss of life expected in the northeast after an NMSZ catastrophic quake.

17. The U.S. Department of Transportation should immediately develop and practice contingency plans and subsequent massive drills for the at-risk CGE states to rapidly respond to loss of major transportation modes throughout affected regions during an M8.0 in California, Alaska, or NMSZ series of M7.3 to M8.0 quakes or singular M7.0 or higher quakes in South Carolina and Puerto Rico.

18. The Department of Defense should remove strategic assets from the high-CGE-risk states. Existing military assets, bases, posts, and related organizations should immediately develop contingency plans to continue their assigned missions and also plan to serve as aid providers of essential supplies and equipment, policing actions, and in a host of disaster recovery operations. All nuclear weapons and associated delivery systems should be relocated out of high-risk earthquake zones across the USA.

19. Make certain that the judicial and legislative branches of the government are fully involved with and supportive of this national threat and the need to prepare for it through legislation and funding of requisite programs.

20. All agencies of the federal government should establish and practice preparation and post-CGE plans and exercises appropriate to their mission.

21. The U.S. government should coordinate this CGE preparation planning with international partners and allies.

This book has been the product of a straightforward extrapolation of USA earthquake history to our current period, demographics, political, energy, social, and economic situation. Past geophysical catastrophes during the previous similar period of solar activity (Dalton Minimum, 1793–1830) documented herein took place before the use of electricity, when there were no radios, no TV, no cell phones or phones of any kind, no internet, and no computers that controlled much of the operation of the country. There were no cars and trucks, no airplanes, no trains. Ocean travel in wooden ships was uncomfortable if not dangerous. The steam engine was in its infancy. The first telegraph was still fifty years away (1850s).

Most people got around by boats or ships at sea, by walking long distances or in horse-drawn carts and wagons or riding on the backs of these horses and mules. Importantly, there were no power grids that could suddenly fail, propelling tens of millions of people back into a cavemanlike existence overnight with no skills or resources to deal with such a calamity!

Only those relatively few who remember, much less endured, days or weeks without power from a hurricane or power system malfunction, like that which struck New York City in 2003 or in San Diego for eleven hours back in September 2011, can appreciate living in a city without power from the grid. During these next predicted catastrophes, whole regions encompassing multiple states may be without power for weeks or months if not longer and, more likely than not, in the middle of winter.

Perhaps most importantly, these past events happened during an era (early 1800s) when most Americans were self-sufficient. They raised their own food or had access to the local butcher or baker and could trade for goods without the need for money per se. They were living without support from any but their own family, and of course, all Americans at that time were completely without any government assistance. Even with a fireplace and ample wood to burn, thousands of Americans still froze to death or starved to death during the last solar hibernation, the Dalton minimum. This was also the last period of cold when the New Madrid Seismic Zone broke apart.[30]

What preparations are recommended in this text are far from lunatic conspiracy theories. They are, rather, the result of comprehensive set of validated science, not politicized science. It is the logical outcome of proper and objective analysis of historical events conducted by serious, concerned scientists, researchers, and organizations with proven records of excellence whose only motives are to help people be prepared for the future.

It is now time for our government at all levels, along with the media, to come clean about the geophysical threats we face. The people must be told and given what they need to prepare for the most destructive geophysical era since our country was founded. This call to arms must also be issued with a sense of urgency—some of these historic earthquakes could hit right after this book is published!

In this book, we have done our best to lay out the science behind what we believe to be the greatest natural threats to our kind for the past 200 years and perhaps far longer. The following is what we have discovered and explained in great detail and restated here as a summary assessment of the predicted catastrophic earthquakes in the United States between now and 2045:

Summary Assessment

High Risk Zone Assessment

1. The NMSZ — There is a minimum 80% chance of a catastrophic earthquake with a magnitude of M6.8 to M8.0 or more likely, a series of such earthquakes affecting this eight-state region between 2017 and 2038. This includes substantial collateral economic damage and personal suffering to the northeastern U.S. states and their people.

2. The CSZ — There is at least a 60% probability that the CSZ will rupture at a magnitude of M8.0 or greater during the period from 2017 to 2038, with strongest likelihood being a segment C event.

3. California	There is at least an 80% chance of more than one catastrophic earthquake of at least M6.7+ hitting the state during the period of 2017 to 2031 and an 80% chance of a catastrophic M7.9 or greater quake striking the state during the same period.
4. Alaska	There is a minimum 82% chance of a catastrophic earthquake of M 7.9 or greater striking the state during the period of 2022 to 2040.
5. South Carolina	There is a minimum 83% chance of a large earthquake in the range of M4.2 to M 7.3 striking the state during the period of 2017 to 2045.
6. Puerto Rico	There is an 80% chance of a catastrophic earthquake of M7.5 to M8.0 striking the state at any time from now to 2039.

****

We strongly urge everyone in the high-risk earthquake regions of the United States and the rest of the planet to heed the messages in this book and prepare not next week, not next month, or next year but to start immediately. Put this book down now and start with a sheet of paper and a pencil or a blank page on your word processor. If you reside within one of the high-risk zones covered in this book, start writing your family preparation plan, including you family protection plan now! Then start to practice the plan.

If you are in a relatively safe, stable geological area of the USA and have friends or family in these high-risk zones, start another plan. This plan will be how to help them. For they may call you or show up at your door unexpectedly if they believe you are their only source of help after they have lost everything. They and your neighbors will come to your house when they have no money, no food, or drinkable water. They will show up if they believe you can provide what they have not or if you have the means of protecting them or their family from the hooligans who have taken over the streets in their town.

Is this admonishment also a bit over the top? Think about it for a few minutes.

If you live in the northeast United States and will be without oil, gas, and power in the middle of winter because the NMSZ has had another M7.3 to M8.0 series of earthquakes, what will you do especially if you and your neighbors haven't planned for such a likely contingency? How will you get through the worst natural catastrophes in U.S. history—ones that cannot be fixed in days or weeks or months?

Again, the window of highest risk opens shortly after this book is scheduled to be published. It comes a full ten years after revelations of the changes about to take place in the Sun to include a dangerous cold climate and the associated historic earthquakes and volcanic eruptions. Predictions about the change in the Sun have been verified. The strong correlation of the next solar minimum or solar hibernation and catastrophic earthquakes has been established in this book.

The scale of the economic damage and recovery costs in the hundreds of billions of dollars to include some states going bankrupt could be too much for our country which is already over $19 trillion in debt. That prospect adds another dimension on top of the natural adversity we will have to face.

Like *Dark Winter*, this book is yet another effort to give our people some means of understanding what our future may be like and how best to prepare for it. It is so very difficult for us to come to grips with the high probability that our generation and those to immediately follow are about to live through the most tumultuous period of natural catastrophes in the last few hundred years, yet that is exactly what the most reliable climate science and history says is about to happen. It is also what some of the best seismologists say is about to happen.

Even with these warnings outlined herein, most Americans will unfortunately do absolutely nothing to prepare. It is human nature. In modern Western civilized societies like that of the USA, we cannot accept that we can be subject to an utterly destructive, lasting, widespread natural disaster. Such rare happenings take place only in faraway lands like Bangladesh or Indonesia or Japan or China, where the U.S. populace and therefore our press have little interest or day to day coverage. People in these countries for hundreds and thousands of years have lived with horrific natural disasters that kill 50,000 or more people at one time on a recurring basis. That has not been so in the USA—until now!

Here in the USA, we have no historical and cultural awareness or fear of these natural cataclysms. There has never been a threat like this during our lifetime or that of our fathers or grandfathers or great-grandfathers. Therefore, most of us will assume we are probably safe.

Most of us will believe to our great misfortune that the state or federal government will come to our aid regardless of what natural calamity befalls us, our city, our state, or even our region of the United States.

In fact, our leaders at the DHS, DOI, FEMA, the USGS, and the White House say that have it under control. They say if you can do a "duck, cover, and grab hold" exercise in an elementary school, high school, or work site, then you will be fine—no worries. As we found in the FEMA response to our letters warning of expected catastrophic earthquakes, our fellow citizens are now "better prepared to respond to these disasters."

Are you?

Appendix 1

The Relational Cycle Theory of Climate Change

(The RC Theory)

1. There exists a family of solar activity cycles that has a profound and direct influence on the Earth's climate.

2. These cycles are called relational cycles since their effects can be experienced, or related to, during one or two human lifetimes.

3. There is a centennial cycle of 90–100 years' duration, which manifests itself with solar activity minimums and associated low temperatures with episodes lasting a few years to 1–2 decades.

4. There is a bicentennial cycle of about 206 years that is the most powerful of the relational cycles and has significant effects on the climate of the Earth lasting many decades, resulting in the most extreme variations in solar activity and in the Earth's temperatures.

5. These cycles are correlated strongly to all past major temperature lows.

6. There is remarkable regularity and hence the predictability of these oscillations such that the theory may be a powerful tool in forecasting of major temperature and climatic cycles on Earth many decades in advance.

7. There may be other relational cycles of shorter duration accounting for lesser solar and climatic events which may be revealed in subsequent research.

Appendix 2

Research Reports, Papers, Etc.

Outline of Items Contained in this Appendix

1. Correlation of Solar Activity Minimums and Large Magnitude Geophysical Events, SSRC Research Report 1-2010 (Preliminary) by John Casey

2. Earthquakes and Solar Activity Cycle, by Choi and Maslov in the NCGT Newsletter No. 57, December 2010

1. Earthquake/Volcanic Activities and Solar Cycles, a paper/commentary by Dr. Dong Choi in the SSRC GCSR, September 13, 2013

2. Earthquakes and Solar Cycles: Increased Earth Core Activity Since 1990,, paper by Casey, Choi, Maslov, and Tsunoda in the SSRC GCSR, March 2014

3. The New Madrid Seismic Zone, Central USA: The Great 1811–1812 Earthquakes, Their Relationship to Solar Cycles and Tectonic Settings, SSRC GCSR paper by Dr. Dong Choi and John Casey

4. USGS History of the 1811–1812 NMSZ Earthquakes

5. List of Researchers Who Predict a New Cold Climate or Solar Hibernation

6. Chronology of Notifications to Government, Media, and the Public of New Cold Climate and Associated Catastrophic Earthquakes and Volcanic Eruptions

7. Earthquake Prediction Center Ends Successful Test Program Early, IEVPC Press Release, December 18, 2012

8. Seismo-Volcanic Energy Propagation Trends in the Aleutian Islands and North America by Dong Choi, Fumio Tsunoda, and Leo Maslov

9. Seismo-Volcanic Energy Propagation Trends in Central America and Their Relationship to Solar Cycles by Dong R. Choi

Appendix 2, Item 1

Correlation of Solar Activity Minimums and Large Magnitude Geophysical Events

SSRC Research Report 1-2010
by John L. Casey

*This has been released for worldwide web distribution on March 1, 1010.

Research Report 1-2010 (Preliminary)

[1] An independent review of historical records was performed for 350 years of global volcanic activity(1650–2009) and seismic (earthquake) activity for the past 300 years (1700 to 2009) within the continental United States and then compared to the Sun's record of sunspots as a measure of solar activity. All three data STs were examined to determine whether a relationship existed between them and if the results of such a study could be used to develop methodology for identifying future geophysical events. The preliminary results from the study have shown that there exists a strong correlation between the solar activity that causes climate changes and the Earth's largest seismic and volcanic events. The impressive degree of correlation for global volcanic activity (>80.6%) and for the largest USA earthquakes (100% of the top 7 most powerful) vs. solar activity lows provides a basis for future estimates of the time periods and magnitudes for the largest volcanic and seismic events many decades in advance. Finally, the coincidence of the Centennial and Bi-Centennial cycles of the RC Theory showed unmistakable relationships to these largest geophysical events. The use of such a tool may provide a new and valuable method for protection of people and property located in and around high risk

geologic zones. Further, a significantly increased risk is indicated during the next 20 years for volcanic and earthquake events of historic scale. Citation: Casey, John. L. (2010), Correlation of Solar Activity Minimums and Large Magnitude Geophysical Events, Research Report 1-2010 (Preliminary), March 1, 2010, Space and Science Research Center, (SSRC).

1. **Introduction.**

[2] Previous work by Casey (2008) known as the "RC Theory," established solar activity as a reliable model for prediction of the Earth's climate changes. During the course of the research it was observed that there may be a positive correlation between solar activity as measured by sunspot counts over a long term base line average, and major geophysical events specifically earthquakes and volcanic eruptions. This previous research found for example, that the largest ever recorded volcanic eruption, Mt. Tambora in Indonesia (1815), as well as the largest earthquakes in the history of the United States, the New Madrid earthquakes of 1811–1812, all occurred near the bottom of the last solar hibernation known as the Dalton Minimum (1793–1830). Given this initial relationship, a more detailed study of geophysical records was made to assess the degree of correlation if any that may exist between the Sun's activity and such events.

2. **Review of Global Volcanic Activity vs. Solar Activity from 1600AD to 2009AD.**

[3] Using the historical record of volcanic eruptions developed by the Smithsonian Institution, an extraction was made of all those eruptions that were rated at a Volcano Explosive Index (VEI) of "5" or greater. The level of 5 on the VEI scale of 0–8 was selected since it was the beginning class of large eruptions. Many are familiar with the scale of such via Mt. St. Helens. This volcano has an established eruptive history and up to and after the May 18, 1980 VEI 5 event, was well documented and instrumented.

---

1. Space and Science Research Center, a division of Verity Mgmt. Svcs. Inc., Orlando, Florida

****

Table 1.Volcanoes of greater than or equal to VEI of 5 from 1650 to 2009. This list of large volcanic eruptions since 1650 was used as the baseline list for comparison against solar activity, i.e. periods of reduced sunspot count to determine any apparent associations. 5* = a class five VEI with potentially large date uncertainty, P* = plinian large class eruption, assumed >VEI 5. The study did not include activity associated with geological hot spots or caldera (super volcano) sites. Source: Smithsonian Institute.

Volcano	Location	Year	VEI
Shiveluch	Kamchatka Peninsula	1650	5
Long Island	N.E. New Guinea	1660	6
Usu	Hokkaido, Japan	1663	5
Shikotsu	Hokkaido, Japan	1667	5
Gamkonora	Halmahera, Indonesia	1673	5*
Tongkoko	Sulawesi, Indonesia	1680	5*
Fuji	Honshu, Japan	1707	5
Katla	So. Iceland	1721	5*
Shikotsu	Hokkaido, Japan	1739	5
Katla	So. Iceland	1755	5
Pago	New Britain	1800	P**
St. Helens	Washington State, USA	1800	5
Tambora	Lesser Sunda Islands, Indo.	1815	7
Galungung	Java, Indonesia	1822	5
Cosiguina	Nicaragua	1835	5
Shiveluch	Kamchatka Peninsula	1854	5
Askja	N. E. Iceland	1875	5
Krakatau	Indonesia	1883	6
Okataina	New Zealand	1886	5

Santa Maria	Guatemala	1902	6
Lolobau	New Britain	1905	P*
Ksudach	Kamchatka Peninsula	1907	5
Novarupta	Alaska Peninsula	1912	6
Cerro Azul	Chile	1932	5+
Kharimkotan	Kuril Islands	1933	5
Bezimianny	Kamchatka Peninsula	1956	5
Agung	Lesser Sunda Islands, Indonesia	1963	5
St. Helens	Washington State, USA	1980	5
El Chichon	Mexico	1982	5
Pinatubo	Philippines	1991	6
Cerro Hudson	So. Chile	1991	5+

****

Of the 31 eruptions documented since 1650 with a VEI greater than or equal to 5, a total of 25occurred during a reduced period of sunspots if not a major reduction in sunspots or a solar hibernation, e.g. the Dalton or Maunder Minimums. This preliminary study showed 80.6% of the largest eruptions took place during extended solar activity minimums. Significantly, the following list of the eight largest volcanic eruptions globally (VEI>6) since 1650, shows all but one took place only during a solar hibernation, or significant reduction in solar activity as measured by sunspot count.

****

Table 2.Volcanic eruptions that took place during major solar minimums and solar hibernations. This table establishes the strong relationship between the largest volcanic eruptions and solar activity lows on the order of the Centennial and Bi-Centennial Cycles defined by the RC Theory.

Volcano	Location	Year	VEI	Associated Solar Minimum
Long Island	N. E. New Guinea	1660	6	Centennial: Maunder
Pago*	New Britain	1800	P	Bi-Centennial: Dalton
Tambora	Lesser Sunda Islands	1815	7	Bi-Centennial: Dalton Indonesia
Krakatau	Indonesia	1883	6	Centennial: Year 1900
Santa Maria	Guatemala	1902	6	Centennial: Year 1900
Lobobau	New Britain	1905	P	Centennial: Year 1900
Novarupta	Alaska Peninsula	1907	6	Centennial: Year 1900
Pinatubo	Philippines	1991	6	No Correlation

****

* P = plinian level large eruption.

3. **Solar activity time line for comparison with volcanic and earthquake activity.** [4] Extended sunspot minimums, i.e., covering two or more 11 year Schwabe cycles, intermediate minimums, and solar hibernations were extracted from the chart below for use in volcanic and earthquake vs. solar activity comparisons:

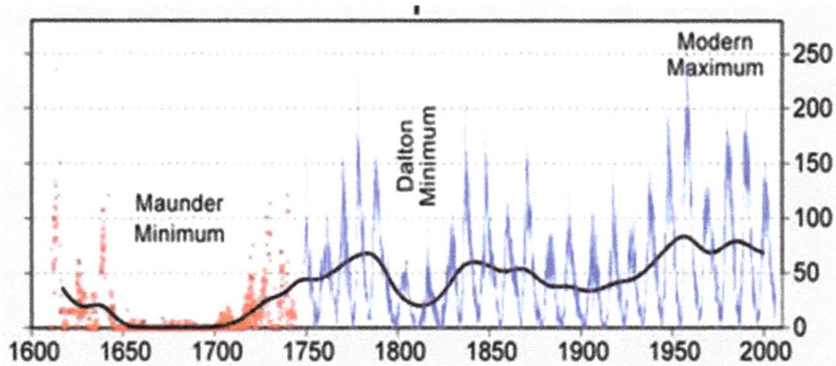

Figure 1. 400 years of sunspot observations depicting the Maunder and Dalton Minimums and the Modern Maximum. Source: Graphics; R. A. Rhode, Global Warming Art, from data STs by Hoyt and Schatten (1998a, 1998b).

4. **Correlation of the Largest Continental Earthquakes to Solar Activity for the Period 1700 to 2009.**

[5]Identification of the largest continental US earthquakes was done with data from the US Geological Survey (USGS). The table below shows the top seven largest earthquakes and is taken from the revised June 7, 2005 published list of the largest fifteen earthquakes and reexamined for completeness on September 8, 2009. All of these largest seven are strongly correlated to an associated solar activity minimum.

****

Table 3. Top seven largest continental USA earthquakes. Source: USGS*

Location	Date	Magnitude	Associated Solar Minimum
Cascadia subduction zone	01-26-1700	~9	Centennial: Maunder
New Madrid, Missouri	12-16-1811	8.1	Bi-Centennial: Dalton
New Madrid, Missouri	02-7-1812	~8 *	Bi-Centennial: Dalton

Fort Tejon, California	01-09-1857	7.9	Intermediate Minimum***
San Francisco, California	04-18-1906	7.8	Centennial Minimum
Imperial Valley, California	02-24-1892	7.8	Centennial Minimum
New Madrid, Missouri	01-23-1812	7.8	Bi-Centennial: Dalton

****

* Measurement methods vary. The USGS says the New Madrid Feb 7, 1812 temblor may have been 8.8 on the Richter scale.

** Centennial and Bi-Centennial cycles from the RC Theory have periods of 90–100 years and 206 years respectively.

*** Intermediate Minimums are easily observed declines in solar activity (sunspots) though lesser in magnitude than Centennial or Bi-Centennial events.

5. **Conclusions.**

[6] As a result of research conducted, it is reasonable to conclude there exists a strong correlation between global volcanic activity among the largest of classes of eruptions and solar activity lows. With the 80.6% occurrence of large scale global volcanic eruptions taking place (>VEI 5) during solar activity lows and with 87.5% occurring for the very largest (>VEI 6) eruptions during major solar minimums, it is concluded that any reliable predictive tool for forecasting future solar activity would also lend itself to forecasts for future global volcanic eruptions of the most powerful magnitudes. For example the RC Theory of solar activity may be an effective tool for forecast of global volcanism.

[7] The occurrence of each of the largest seven USA earthquakes during solar activity lows and in particular during solar hibernations indicates a predictive tool like the RC Theory for future extended solar minimums may also be effective in forecasts of major USA earthquakes.

[8] Given the unusually high degree of correlation found in the study for both the highest levels of global volcanism and USA earthquake activity when compared to extended solar activity lows, it can be concluded that there exists a significant likelihood (greater than 80%) that the current recently started solar hibernation may result in historic scale global volcanic eruptions and record earthquake activity within the continental United States.

[9] The determination that solar activity cycles may indicate timing and intensity of geophysical events like volcanism and earthquakes points toward a possible connection between solar activity and the underlying cause of these geophysical events, namely plate tectonics.

[10] The solar hibernation identified by Casey (2008) is currently under way. The results of this study and the high correlation between described volcanism and earthquakes and solar hibernations warrants the widest dissemination of warnings to personnel and governing organizations in high risk geophysical zones. It is expected beginning at any time and during the next twenty years of the solar hibernation, that potentially historic volcanic eruptions are likely globally and similarly record setting new earthquakes are likely within the continental United States.

[11] Acknowledgements.

The Global Warming Art project and the GNU Free Document Licensing program in total, for their open availability of high fidelity graphical representation of important research data, especially by R. A. Rhode and L. McInnes and data from Hoyt and Schatten. The US Geological Survey for historical data on USA earthquakes. The Smithsonian Institute for data on global volcanism.

References

Siebert L., Simkin T (2002–). Volcanoes of the World: an Illustrated catalog of Holocene Volcanoes and their Eruptions. Smithsonian Institution, Global Volcanism Program Digital Information Series, GVP-3, (http:www.volcano.si.edu/world/).

USGS, Largest Earthquakes in the United States, June 07, 2005, http:www.earthquake.usgs.gov/regional/states/10_largest_us.php.

Casey, John L. (2008) The existence of 'relational cycles' of solar activity on a multi-decadal to centennial scale, as significant models of climate change on Earth, Research Report 1-2008, January 22, 2008, Space and Science Research Center, (SSRC).

Appendix 2, Item 2

Earthquakes and Solar Activity Cycles

Dong R. Choi
Raax Australia Pty Ltd
raax@ozemail.com.au; www.raax.com.au

Leo Maslov
Aims College, Greeley, CO, USA/
Ms_leo@hotmail.com

Abstract: Earthquake occurrence from 1973 to 2010 shows that earthquake frequency is closely related to the solar cycle: the number of earthquakes increases during the declining/trough periods. A small sharp spike occurs at the peak. On the other hand, earthquakes decrease during the rising period of the solar cycle. Three earthquake cycles are recognized: 11, 22 and 44 years. **1)** The 11-year solar (sunspot) activity cycle generally affects shallow (300 km or shallower) and small to relatively strong (M6.9 or below) earthquakes. **2)** The 22-year seismicity cycle is best seen in conspicuous low frequency troughs of the M5.0-5.9 quake group at shallow depth. They correspond to solar cycles 21 and 23 in their rising years. **3)** Very strong (M7+) shocks at all depths are related to a longer cycle, 44 years—from 1969 to 2013; a seismically dormant period occupies the first half, whereas an active period occupies the second half of the cycle, which corresponds to the declining years of the 44-year cycle. **4)** The largely amplified seismic trough or period of quiescence from 1996 to 2003 cannot be attributed to solar activity alone. It will need to invoke additional forces—such as planetary interactions. **5)** The rapid increase in small (below M4.9), shallow (100 km or shallower) shocks after 1994 is due to the increased activity of very deep, very strong shocks since 1990. A very high level of seismic and magmatic activities is expected to occur from 2012 to 2014 when the 44-year solar cycle reaches the minimum point which coincides with the peak period of the relatively weak solar cycle no. 24. The years from 2012 to 2014 also coincide with the end (or minimum point) of the 86-year solar cycle (1928-2013).

Keywords: *earthquake cycle, solar cycle, Sun-Earth interaction, 44-year earthquake cycle, seismic quiescence*

Introduction

In recent years the study of solar activity and its effects on the Earth's tectonic activity has dramatically increased: Simpson, 1967; Zhang, 1998; Gregori, 2002; Gousheva et al., 2003; Duma and Ruzhin, 2003; Endersbee, 2007; Khain and Khalilov, 2008; Hissink, 2009; Casey, 2010; Quinn, 2010, to cite only a few. This work has indisputably proved the Sun's pervasive influence through its electromagnetics, flares, protons, gravity, etc. on many geophysical activities of the Earth including earthquakes, volcanic eruptions and global warming.

This paper is an extension of our earthquake study from the viewpoint of global geodynamics, and attempts to correlate seismicity and solar activity based on the seismicity data archived by USGS NEIC from 1973 to 2010 and the solar cycle data extracted from NASA websites.

General Trends in Annual Fluctuation of Earthquakes and Their Correlation with the Solar Cycles

Figure 1 compares the total number of global seismic events recorded in the NEIC archive from 1973 to 2010 (December inclusive) for various categories—depths (shallow depth above 300 km; deep below 300 km) and magnitudes (M4.0 to 4.9, 5.0 to 5.1, 6.0+, and 7+) with the sunspot cycles. **Figures 2 & 3** provide the time-depth plots for various magnitudes.

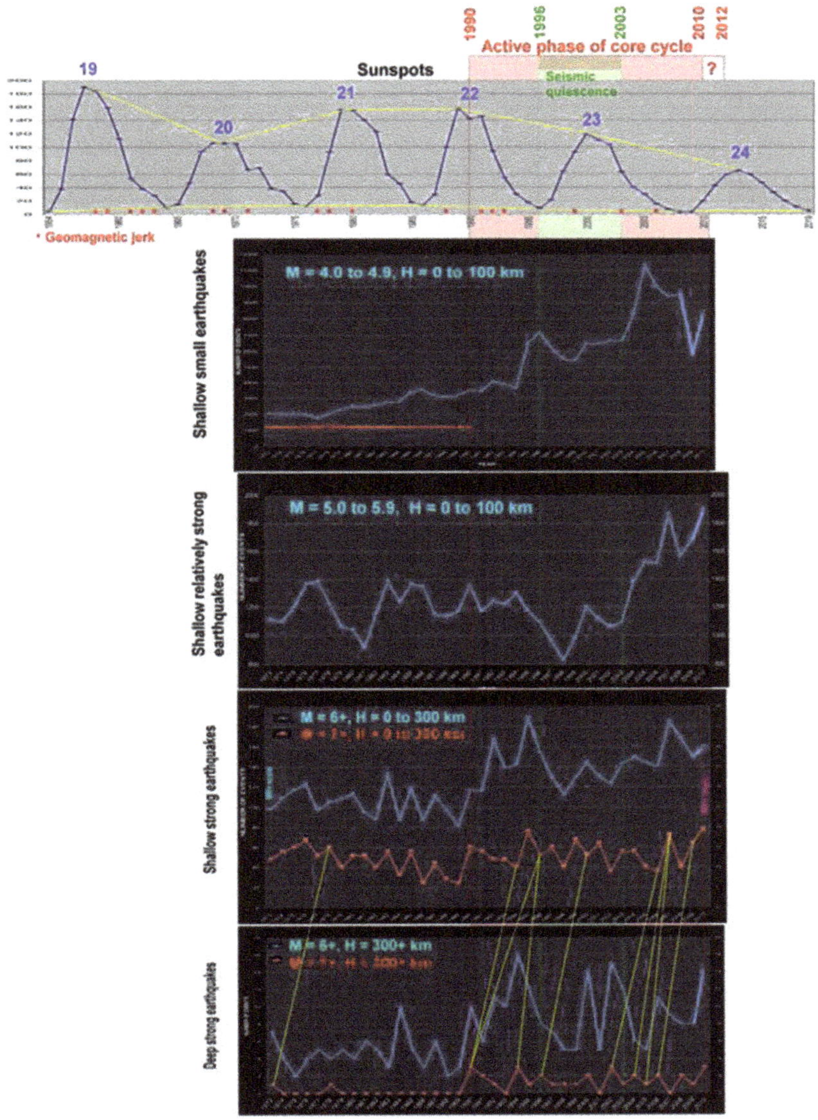

Figure 1. Annual fluctuation of earthquake numbers at varying depths and magnitudes in comparison with the sunspot cycle curve. The yellow lines in the bottom two figures are links established by Blot's ET concept (1976). Sunspot data from NASA (http://solarscience.msfc.nasa.gov/SunspotCycle.shtml; http://solarscience.msfc.nasa.gov/predict.shtml). Geomagnetic jerks at the core-mantle boundary are from Quinn (2010).

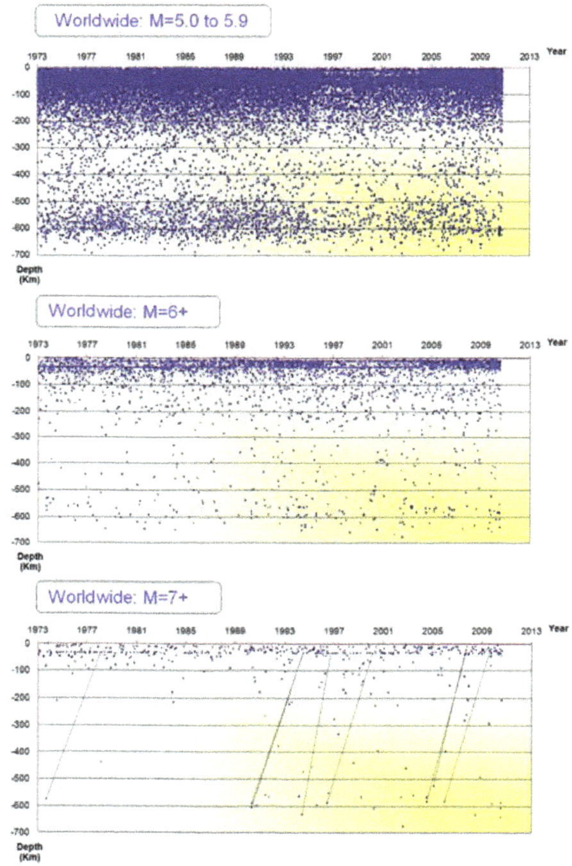

Figure 2. Time-depth plots of all registered quakes with magnitudes M5.0 or greater from 1 January 1973 to 15 August 2010 worldwide. The links between the deep and shallow quakes established by Blot's ET formula are indicated in the bottom figure. Note a quiescent period from 1996 to 2003 observed in quakes below M7.0 in all depths. It is most conspicuous in the M5.0 to 5.9 quake group, and also well expressed in the M4.0–4.9 earthquakes, see Fig. 3. But it is not recognized in the strongest quakes above M7+ in magnitude, particularly in the deep section. The sudden appearance of strong shocks above M7+ from 1990 onwards is most conspicuous. The M6.0 to 6.9 group has also markedly increased from 1990 onwards.

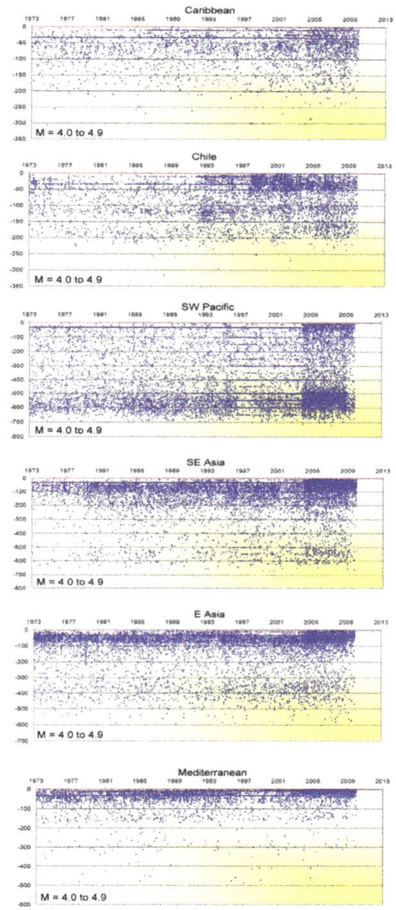

Figure 3. M4.0–4.9 quake time-depth plot from 1973 to 2009 for selected areas. Some artificial influence is seen in the data set of the small quakes of this magnitude: Caribbean until 1987, Chile 1991, and SE Asia 1978. See Choi (2010a) for additional information. The seismic quiescence in both shallow and deep sections from 1996 to 2003 is unequivocally expressed in these figures.

Three prominent trends are recognized in these graphs: 1) shallow (above 300 km) small (M4.0 -4.9) shocks showing a general increase at least from 1995 onwards and a general match with the 11-year solar cycle, 2) relatively strong (M5.0 to 5.9) quakes showing a correlation with the 11-year and 22-year solar cycles, and 3) strong (M6.0 or greater) shocks affected by a longer solar cycle (44 years).

1) Small magnitude group, M4.0 to 4.9

The small shocks in this range account for almost 80% of the registered quakes with magnitudes 4.0 or greater. Setting aside the pre-1990 data due to the possible data bias, there is an overall rise from 1991 to 2010: a slow rise from 1991 to 1994 followed by a sudden jump from 1995 to 1996. After a small depression and stability from 1997 to 2003 (global seismic quiescence), the curve again rises rapidly with peaks from 2005 to 2008. This trend has remained high in 2010 too.

The sharp spike in 1995–1996 corresponds to the solar trough between cycles 22 and 23, and the following large peak from 2005 to 2008 to the descending phase of the solar cycle 23.

The sudden flurry of small, shallow shocks from 1995 is due to the arrival of energy derived from strong (M7+) deep shocks which occurred 3 to 5 years prior to the shallow quakes; the links between the deep and shallow very strong shocks (M7+) established by the energy transmigration concept (Blot, 1976; Choi, 2010c; Table 1) are indicated in Figs. 1 and 2.

Year	Month	Day	Latitude	Longitude	Magnitude	Deph (km)	Note
1973	9	29	41.89	130.87	7	575	Linked to: 23 & 24 Mar/1978. M=7.5-7.6, H=33km. Offshsore Iturup, Russian Far East. New data.
1990	5	12	49.04	141.85	7.2	605	Linked to: 4 & 9 October, 1994, M=8.3 and 7.3, H = 14 and 33 km, respectively. Offshore Iturup, Russian Far East. New data.
1990	5	24	-7.36	120.36	7.1	588	Linked to: 21 February, 1996, M=7.5, H=10 km, Offshore Peru. New data.
1994	6	9	-13.84	-67.55	8.2	631	Linked to 12 Nov. 1996 quake. H=33, M=7.7. 350 km South of Lima (Choi, 2010c)
1996	6	17	-7.14	122.59	7.9	587	Linked to 14 May 2000 quake. M=6.3, H=33. Suva Sea, Indonesia (Choi, 2010c)
2002	8	19	-21.7	-179.51	7.7	580	Linked to 5 May, 2006 M=8.0 quakes (H=11km), 100km north of Tonga (Choi, 2010c)
2002	8	19	-23.88	178.49	7.7	675	Ibid. (H= about 20 km)
2004	7	25	-2.43	103.98	7.3	582	Linked to the Southern Sumatra quake (M=8.4) on 12 Sep. 2007 (Blot and Choi, 2007)
2005	2	5	5.29	123.34	7.1	525	Northern Celebes Sea. Possible link: 26 July, 2007. M6.9, H 25 km (best match 35 km). Molucca Sea (Choi, 2010c).
2006	1	2	-19.93	-178.18	7.2	582	One of 29 Sep. 2009 Samoa quake forerunners (Choi, 2010a)

Table 1. Some of the deep (500+ km) very strong (M7+) shocks and their shallow links established by the ET concept. See also Choi (2010c) for more deep strong shocks.

The seismicity rise from 1992 to 1996 correlates with the declining period of solar cycle 22, the overall depression in seismicity from 1997 to 2003 correlates with solar cycle 23's rising period, and the seismicity peak from 2004 to 2008 correlates with the declining period of solar cycle 23. The M4.0–4.9 quakes are correlated with the 11–year solar cycle.

2) Relatively strong group, M5.0 to 5.9

The shocks in this group show highs in 1976–1977, 1983–1986, and 2004–2010 (Fig. 1, central figure, and Fig. 2 top), all of which correspond to the declining years and troughs of the 11-year solar cycles nos. 20, 21 and 23, respectively.

Regarding the low frequency periods, there are three troughs: 1) from 1978–1982 (five years), 2) from 1987 to 1989 (three years), and 3) from 1996 to 2003 (eight years). The last one is most conspicuous and best illustrated in the time-depth plot graphs, Figs. 2 & 3,in smaller quakes below M5.9, but some impact is also seen in the M6.0-6.9 quakes. The sudden drop from 1996 to 1998 is particularly dramatic—about a 40% reduction in number from the peak; the dramatic reduction is also seen in all other quakes with magnitudes below M7.0. This particular period covers the first two thirds of solar cycle 23, from the rising period, through the peak, to the first half of the declining period. It is three to five years longer than other troughs

It is interesting to note that seismicity drops when the solar cycle starts to rise and keeps falling or stays low until the solar cycle starts to decline after passing the peak. The falling seismicity trend reverses toward the rise at the middle of the declining period of the solar cycle. This is especially well observed in the M5.0-5.9 graph, Fig. 1, in solar cycles 21 and 23. However, it is not well expressed in solar cycle 22. Therefore, the deep troughs are considered to have a cycle of 22 years. Also noted is the sharp (single-year), small rise at the peak solar year—which is observed in 1980, 1990 and 2000 in most of the quakes.

3) Strong shocks, M6.0 or greater

The fluctuation of the M6.0+ group is illustrated in the bottom two figures of Fig. 1 and their time-depth plot in Fig. 2.

The fluctuation curve of the M6.0+ shocks is different from the smaller magnitude groups below M6.0. In both the shallow and deep groups belonging to the M6.0+ group, two distinctive periods are recognized: low from 1973 to 1989, and high from 1990 to 2010. This is particularly clear in the very strong (M7.0+) deep (300 km+) group, which is not affected either by the quiescent period from 1996 to 2003. This pattern is related to neither the 11-year nor the 22-year cycle. We consider that the seismicity frequency curve observed in M6.0+ is primarily related to the 44-year cycle, which is seen in the long-term solar frequency curve, from 1969 to 2013, covering the peaks of solar cycles 20 to 24 (Fig. 1). 1990 is the peak year of this 44-year cycle;

the active phase of the deep M7+ seismicity falls in the declining year of this longer cycle. A strong seismicity peak in 1990 coincides with the 44-year solar peak. The 11-year solar cycle peak also accompanies a small seismicity spike, as stated earlier.

However, some of the M6.0+ quakes are affected by the 11-year solar cycle: both the shallow and deep M6.0–6.9 quakes show a marked drop from 1996 to 1998—the first half of the rising period of solar cycle no. 23. This is also the first half of the most prominent seismic quiescent period from 1996 to 2003, implying that an extraordinarily strong force (deep penetration and in magnitude) had been in operation which caused the prolonged dormancy of the Earth's tectonic activities.

Discussion

The current study reveals many intriguing facts implying close Sun-Earth interactions with cycles of 11, 22 and 44 years, together with a conspicuous seismic quiescent period. In addition, it is seen an anti-correlation between number of sunspots and number of earthquakes, with a small increase in quakes at solar maximum. This relation is diagrammatically shown in Fig. 4.

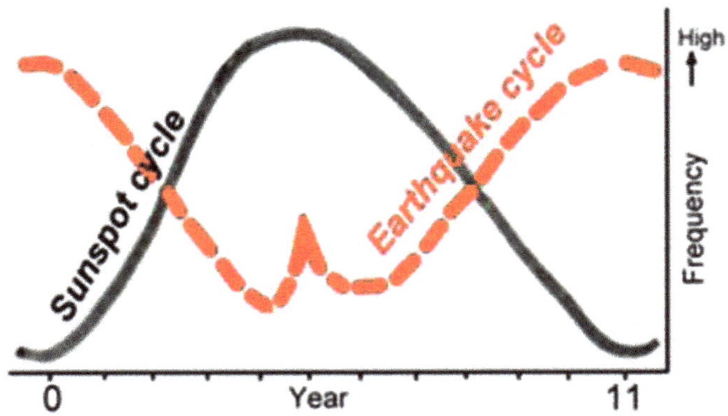

Figure 4. Schematic diagram showing the relation between the solar and the earthquake cycles seen in the 11-year cycle. The trend is applicable for the 22- and 44-year cycles too.

These trends are applicable to volcanic activities too. Casey (2010), who studied volcanic records from the last 350 years and seismicity in the last 300 years within the continental United States, found a strong correlation between solar activity, the largest volcanic eruptions and the strongest earthquakes; the latter two have occurred during the solar activity lows with the strongest ones during the major solar minima. However, another study by Gregory (2002) who compared the historical volcanic eruptions especially Mt. Etna and Vesuvius since 1550 AD and the solar cycles found synchronous relation between them. On the other hand, Quinn (2010) pointed out that the significant volcanic events tend to cluster around geomagnetic jerk events which occur at the core-mantle boundary caused by electromagnetic induction from interplanetary magnetic field and solar wind.

1) The 11- and 22-year cycles

Many previous studies clarified that earthquake activity has cycles about 11 and 22 years like the solar activity cycles (Luo and Li, 1978, cited in Zhang, 1998). The present study supports the observation made by Gousheva et al. (2003): earthquakes occur frequently around the descending phase of solar activity, but relatively less during the peak years of solar activity

Another earlier study on earthquakes and solar activity cycle by Simpson (1967) based on the data from 1950 to 1963 for shocks with magnitude 5.5 or greater also shows similar results: the maximum quake frequency occurs during the period in which solar activity is declining and changing—"solar activity, specifically solar flares and associated magnetic storms do exert a triggering effect on earthquakes." Also noteworthy here is that the period of major flare incidence is during the intermediate years of increasing or decreasing solar activity, with the highest flare incidence occurring during the declining years of the 11-year solar cycle (Das Gupta and Basu, 1965).

A study by Khain and Khalilov (2008), who examined historical data from 1600 to 2000, also found 11- and 22-year cycles in volcanic and seismic activities. But the strong quakes with M7+ stood out most in the 22-year and the volcanic cycle in the 22 to 23-year harmonics. In the solar fluctuation curve the 22-year cycle is not seen in recent years, but three 22-year cycles are recognizable from 1862 to 1928 (Fig. 5).

Gousheva et al. (2003) recognized two maxima in earthquake occurrence corresponding to: 1) sunspot maxima, and 2) the descending phase of solar activity. These observations are fully supported by our data (Fig. 4). The former is usually expressed by a single-year spike, and the latter by a 4 to 6-year rising trend. The seismicity spike is also seen in the 44-year solar cycle peak year—1990.

2) The 44-year cycle

As noted earlier, there is a distinctive trend in deep, very strong (M7+) quake occurrence: consistent appearance from 1990 to the present day, 2010. Before 1990 almost no deep, very strong seismicity had taken place (Figs. 1 & 2). This trend, quiet from 1974 to 1989 and active from 1990 to 2010, is also observed in the M6.0–6.9 category in both deep and shallow depths, although influenced to some extent by the 11-year cycle. However, it is not seen in smaller quakes below M6.0—whose fluctuation cycles are harmonious with the 11- and 22-year solar cycles.

To understand this longer-term, deep, strong quake pattern, we compiled the NASS's historical sunspot data since 1955 and its projections until 2020. This work revealed the existence of a longer-term sunspot cycle: 44 years from 1970 to 2013 with a peak in 1990 (Figs. 1 & 5). This longer cycle is also observed from 1928 to 1969, which is 42 years in duration, and peaked in 1947. Although the 44-year cycle is not clearly observable from 1804 to 1928, it is again seen in the earlier record, from 1761 to 1804, which is a 44-year span too.

We consider this 44-year solar cycle to be responsible for the trend seen in the deep, very strong shocks. The year 1990 coincides with the peak year of the contemporary 44-year cycle (Fig. 1)—meaning that the active phase of deep, very strong seismicity correlates with the declining period of the current 44-year cycle, the same pattern seen in the 11-year cycle.

No previous studies have positively shown the presence of the 44-year seismicity cycle. We believe this was caused by the way earthquake data have been handled. In the previous studies both deep and shallow earthquakes were treated collectively as one set of data; no separation was made between deep and shallow quakes, nor was any consideration given to the genetic links between them.

We will have to wait and see how long the active phase of the 44-year cycle represented by the M7+ deep shocks will last in the coming years. But on the basis of the Tsunoda (2010) study, a 30 to 50-year volcanic-earthquake cycle (VE process) in Asia, which was supported by Quinn (2010), 30 to 60 years on a global scale, as well as the above discussion, we can expect the activity to head for a dormant period from 2013 onwards when the 44-year sunspot cycle enters the next cycle, while the shallow strong quakes continue for an additional 4 to 5 years.

In addition to the 44-year earthquake cycle, Duma and Vilardo (1998) and Gousheva et al. (2003) showed the possible presence of an 80-year earthquake cycle. The former analyzed worldwide earthquakes with magnitude 7 or greater, and the latter M3.1 to 5.0 in Austria from 1893 to 1992 (their figure 6). The latter also showed that the curve of magnetic intensity matched earthquake fluctuations. An 86-year solar cycle from 1928 to 2013 is also recognizable in Fig. 5. However, as we have no comprehensive seismicity data set for the period prior to 1973, it is impossible to develop any meaningful discussion here. Other than these cycles, Quinn considers the most prominent cycle in Earth magnetism to be the 60-year cycle, while the 30-year quasi-period is related to very strong seismicity (M7+) (personal communication, 12 November, 2010).

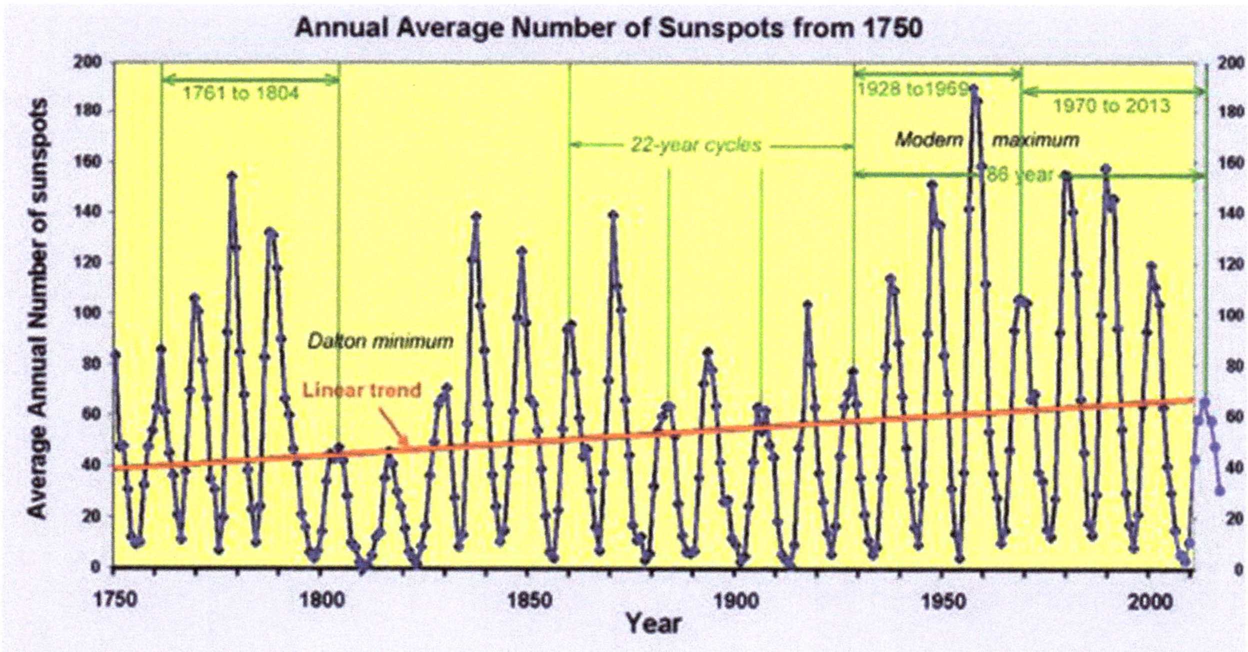

Figure 5. Sunspot cycles compiled by Endersbee (2007) and from the NASA websites. The 22- and 44-year cycles are added.

Considering that solar cycle 24 is expected to be a weak cycle (Hathaway, 2010) and that the highest seismicity coincides with the solar cycle trough, the end or trough of the 44-year cycle from 2012 to 2014 is expected to become a very active period for tectonic and magmatic activities. The period also coincides with the end of the 86-year solar cycle. Casey (2010) too expressed a similar opinion based on his own observations: the strongest volcanic and seismic activities in the continental USA in the last 300 to 350 years have occurred during the major solar minima. The unprecedentedly strong deep shocks in the northern Celebes Sea in July 2010, which have been predicted to reach shallow Earth in late 2012 in the Molucca Sea (Choi, 2010c), can be interpreted as one of these indicators.

3) Seismic quiescence

The conspicuous 8-year quiescent or dormant period in seismic occurrence from 1996 to 2003 worldwide has been noted by the senior author (Choi, 2010b). The quiescence is recognized in all shocks at all depths with magnitudes below M6.9, especially below M5.0 in the shallow group (Figs. 1–3). However, no quiescence is seen in very strong M7+ quakes at both shallow and deep depths. The period starts from the very beginning of sunspot cycle no. 23 and ends two years after the peak.

Although much smaller in amplitude, a similar period of seismic quiescence is observed from 1977 to 1982 (6 years) in the relatively strong group, M5.0 to 6.9 at shallow depth (Figs. 1 & 2). This period corresponds to the high period of solar cycle 21. Therefore the strong seismicity troughs have a 22-year cycle.

The unusually strong, deep-penetrating and long-lasting 1996-2003 quiescence cannot be explained by the solar cycle anomaly alone. It will need to invoke additional forces, such as planetary alignments, together with the relative (baricenter) motion of Earth, Moon, Sun and other planets and the interaction of such motions with Earth's angular momentum (Quinn, personal communication, 12 November 2010).

4) Correlation coefficient

To scientifically examine the above-mentioned correlations between earthquakes and sunspot fluctuation we calculated the correlation coefficient. The calculation was conducted for the M=5.0 to 5.9 shallow group (depth range from 0 to 100 km) and compared with the sunspot cycle, as shown in Fig. 6. This yielded a negative correlation value of -0.5.

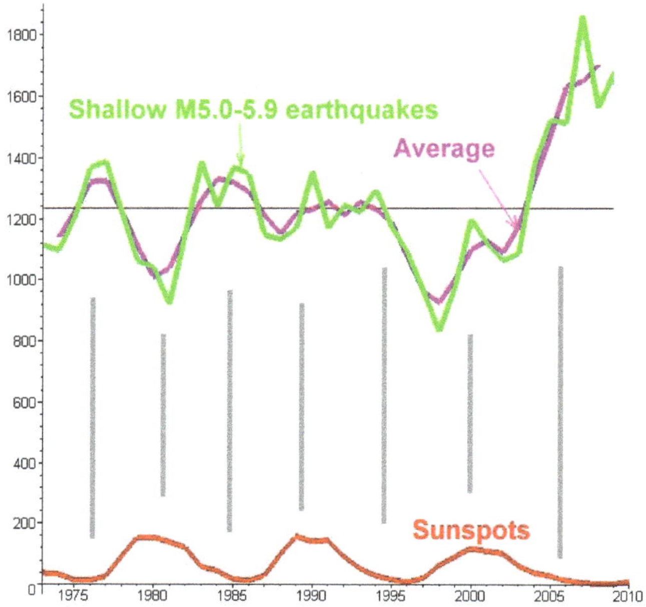

Figure 6. Correlation between earthquakes and sunspot cycle for the shallow (top 100 km), M=5.0 to 5.9 earthquakes from 1973 to 2009. The correlation coefficient is -0.5. Grey lines are added to visually compare the solar and earthquake trends.

5) Possible planetary alignment or interactions which influenced the 1996-2003 quiescence

Variation of a tidal force

If the tidal force the Moon exerts on the Earth is given a value of 1.0, the gravitational force the Sun exerts on the Earth is approximately 0.3, and the gravitational force Jupiter exerts on the Earth is only 0.01. Avsyuk and Maslov (2010) show that the long-period variations of the tidal force in the Earth-Moon-Sun system cause long-period variations of the Earth and Moon seismicity, as well as tides in the Earth's atmosphere and hydrosphere. This force is relatively weak, but it can trigger stresses above some threshold level, accumulated in the Earth due to internal processes. Figure 7 shows the number of earthquakes registered for the period 1973–2009 with magnitudes M = 4.0 to 4.9 at all depths.

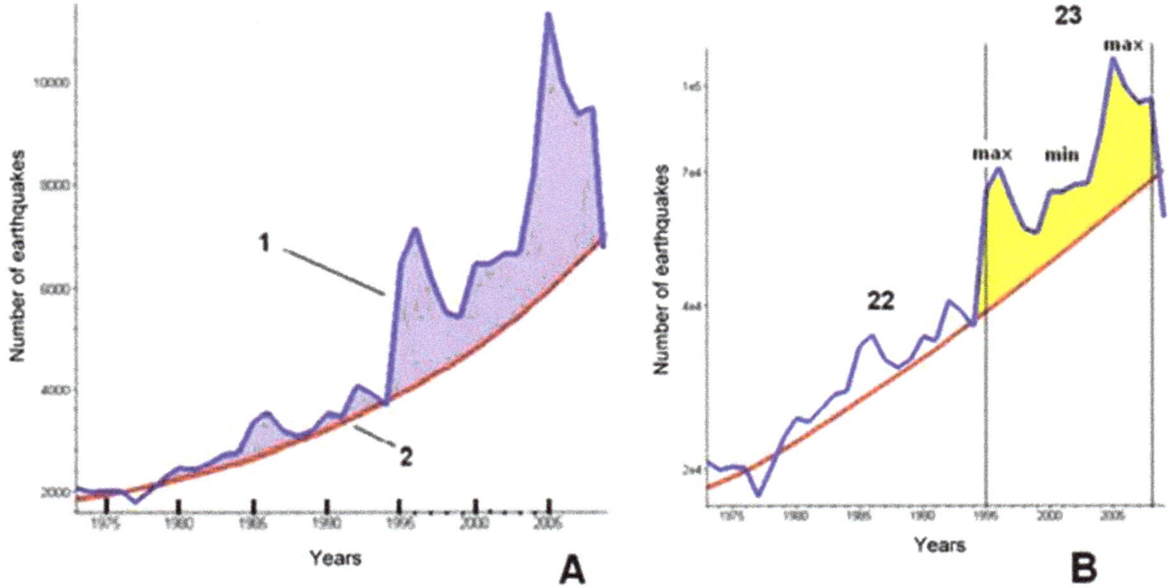

Figure 7. Earthquakes number for period 1973 – 2009 with magnitudes M = 4.0 to 4.9, all depths.

A: 1 – number of earthquakes, 2 – exponential trend.

B: The same data on a logarithmic scale; 22 and 23 are sunspot cycles.

Cycle no. 23 is shown from minimum to minimum of solar activity. Within this cycle we can see two maxima at either end of the cycle and a relative minimum in the middle of it. One possible explanation of this relative minimum (the number of earthquakes is still high, and above the exponential level) is that the rate of release of accumulated stress by triggering earthquakes is higher than the rate of accumulation of stresses in the Earth. Stresses accumulated during the quiescent period are released in the second earthquake peak.

Magnetic field

Jupiter's year is 11.86 earth-years, which is very close to the period of solar activity. Its orbit has a relatively high eccentricity which causes the distance between Jupiter and Sun to vary greatly at 11.86-year intervals. To balance Jupiter, the Sun moves 750,000 km in either direction, causing a total wobble of 1,500,000 km, which is greater than the diameter of the Sun. This process can "stir" the Sun's convective currents and enhance or dampen its activity.

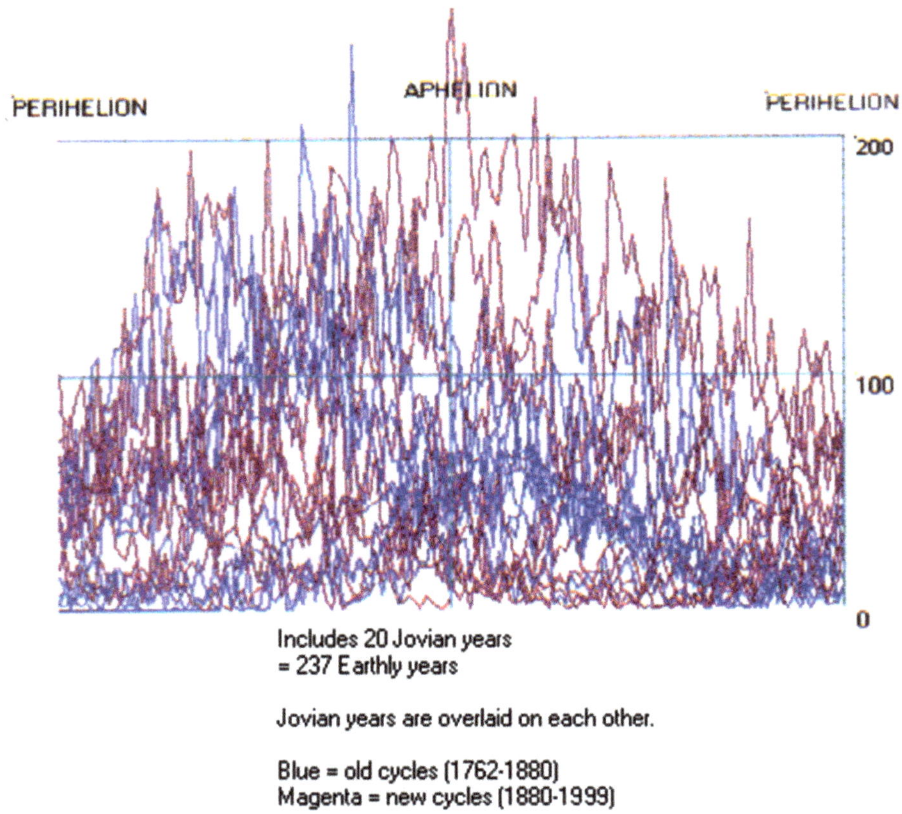

Figure 8. Variation of sunspot numbers during 1772 – 1999 years from perihelion to perihelion of the Jovian orbit (points closest to the Sun).

As seen from the above picture (Fig. 8), when Jupiter is nearperihelion (closest distance to the Sun), sunspot activity is minimal, and when Jupiter is near aphelion (farthest distance from the Sun), sunspot activity is maximal. Jupiter is not the primary cause of the sunspots but it twists the solar magnetic field in ways that affect them.

This figure shows that Jupiter does have a significant effect on the Sun's sunspot activity, which, in turn, affects the Earth's magnetic field and other planetary phenomena.

The mechanisms by which the Sun affects Earth's seismicity are still unknown. We can conclude from our study that it may be a gravitational pull and also an electromagnetic interaction between the Sun, the Earth's geophysical fields (primarily the magnetic field) and Earth's deep structures, especially the electric currents in its outer core.

Conclusions

1. The earthquake number increases during the declining period of the solar cycle. A small sharp spike also occurs at the peak of the solar cycle. Whereas earthquakes decrease in frequency during the rising period of solar activity.

2. The 11-year solar (sunspot) activity cycle generally affects shallow (300 km or less) and small to medium (M6.9 or below) earthquakes. It is noteworthy that the major solar flares occur during the declining years of the 11-year solar cycle.

3. The 22-year cycle is recognized in the low frequency period including the period of seismic quiescence. It coincides with the rising and high periods of sunspot cycle nos. 21 and 23. These cycles are best observed in the shallow, relatively strong quake group, M5.0 to 5.9.

4. The 44-year cycle is well observed in the very strong (M7+) shocks at all depths from 1968 to 2013. This strong group is not affected by the 11-year solar cycle. A seismically dormant period occupies the first half to 1989, whereas an active period occupies the second half of the cycle from 1990 to 2013(?), corresponding to the declining years of the 44-year cycle.

5. The period of seismic quiescence or low frequency period coincides with the rising stage of the solar cycle. It is observed in all magnitudes, especially small and shallow shocks, except for the strongest group, M7.0+. The most pronounced quiescence from 1996 to 2003 is best considered to have been amplified by other external forces – such as planetary alignments and interaction with other planets.

6. The rapid increase in small, shallow shocks (below M4.9, 100 km or shallower) after 1994 is due to the increased activity of very deep, very strong shocks which occurred 3 to 5 years prior to the arrival at shallow depth and triggered a flurry of secondary, smaller shocks.

6. Further study is needed to understand the intricate interaction between solar activity and the Earth's core activity including electromagnetic jerks, and the processes and mechanisms

by which energy is generated and released to the overlying lower mantle and travels to the upper mantle where deep earthquakes and tectonic activity occur.

Acknowledgements: The authors thank John Quinn, Giovanni Gregori and Gerald Duma for their critical review of this paper. Our thanks are extended to David Pratt for his linguistic and editorial comments.

References

Abarcadel Rio, R., Gambis, D., Salstein, D., Nelson, P. and Dai, A., 2003. Solar activity and the earth rotation variability. *Journ. Geodynamics*, v. 36, p. 423–443.

Avsyuk, Yu.N. and Maslov, L.A., 2010. Tides, seismicity and regularities in orbital motion of the Earth-Moon binary planet system. *Earth, Moon and Planets*(in press).

Blot, C., 1976. Volcanisme et séismicité dans les arcs insulaires. Prévision de ces phénomènes. *Géophysique*, v. 13, Orstom, Paris, 206p.

Casey, J.L., 2008. The existence of 'relational cycles' of solar activity on a multi-decadal to centennial scale, as significant models of climate change on Earth. *Space and ScienceResearchCenter, Research Report 1-2008* – The RC Theory. p. 1–8.

Casey, J.L., 2010. Correlation of solar activity minimums and large magnitude geophysical events. *Space and ScienceResearch Center, Research Report 1-2010* (Preliminary), p. 1–5.

Choi, D.R., 2010a. Tectonic significance of the 29 September 2009 Samoa earthquake. *NCGT Newsletter*, no. 54, p. 23–35.

Choi, D.R., 2010b. Global seismic synchronicity. *NCGT Newsletter*, no. 55, p. 66–73.

Choi, D.R., 2010c. Blot's energy transmigration concept applied for forecasting shallow earthquakes: a swarm of strong deep earthquakes in the northern Celebes Sea in July, 2010. *NCGT Newsletter*, no. 56, p. 75–85.

Chown, M., 2004. Chaotic heavens. *New Scientist*, v. 181, issue 2436, p. 32–35.

Das Gupta, M.K. and Basu, D., 1965. Solar-terrestrial events in relation to the phase of the solar cycle. *Jour. Atmosphere, Terr. Phys.*, v. 27, p. 1029.

Das Gupta, M.K. and Sarkar, S.K., 1971. Spectrum of average flux of the S-component of solar radio emission. *Solar Physics*, v. 18, issue 2, p. 276–283. 1971SoPh...18..276D.

Duma, G., 2005. A significant electromagnetic process controls earthquake activity. *Geophysical Research Abstracts*, v. 7, 02860. European Geosciences Union.

Duma, G. and Ruzhin, Y., 2003. Diurnal changes of earthquake activity and geomagnetic Sq-variations. *Nat.*

HazardsEarth Syst. Sci., v. 3, p. 171–177.

Duma, G. and Vilardo, G., 1998. Seismicity cycles in the Mt.Vesuvius area and their relation to solar flux and the variations of the earth's magnetic field. *Phys. Chem. Earth*, v. 23, no. 9-10, p. 927–931.

Endersbee, L., 2007. Climate change is nothing new! *NCGT Newsletter*, no. 42, p. 3–17.

Gousheva, M.N., Georgieva, K.Y., Kirov, B.B and Antanasov, D., 2003. On the relation between solar activity and seismicity. Proceedings of conference on Recent Advances in Space Technologies. RAST '03, p. 236–240.

Gregory, G.P., 2002. Galaxy–Sun–Earth relations. The origin of the magnetic field and of the endogenous energy of the Earth, with implications for volcanism, geodynamic and climate control, and related items of concern for stars, planets, satellite, and other planetary objects. A discussion in a prologue and two parts. *Beiträgezur Geschichte derGeophysik und KosmischenPhysik*, Band 3, Heft 3, 471p.

Haigh, J.D., Winning, A.R., Toumi, R. and Harder, J.W., 2010. An influence of solar spectral variations on

Radioactiveforcing of climate.*Nature,* v. 467, p. 696-699 (7 October 2010). Doi:10.1038/nature09426.

Harris, C. and Mann, R., 2010. Global temperature trends from 2500 BC to 2040 AD.

www.longrangeweather.com/global_temperatures.htm.

Hathaway, D.H, 2010. Does the current minimum validate (or invalidate) cycle prediction methods? ASP Conference Series, arXiv:1003.4208v1 [astro-ph.SR].

Hissink, L.A.G., 2009. The Earth in an electric solar system. *NCGT Newsletter*, no. 51, p. 23–34.

Khain, V.E. and Khalilov, E.N., 2008. About possible influence of solar activity upon seismic and volcanic activities: Long-term forecast. *Science Without Borders. Transactions of the International Academy of Science H & E.*, v. 3, p. 217–240.

Luo, B.R. and Li, W.B., 1978. The stability of sunspot cycle reflected by the records of Chinese ancient aurora and the earthquakes. *Chinese Science Bulletin*, v. 23, p. 8–13 (in Chinese).

NASA, MarshallSpaceFlightCenter, 2010. The sunspot cycle. http://solarscience.msfc.nasa.gov/SunspotCycle.shtml.

NASA, MarshallFlightCenter, 2010. Solar cycle prediction. http://solarscience.msfc.nasa.gov/predict.shtml.

Quinn, J.M., 2010. *Global warming*. Geophysical counterpoints to the enhanced greenhouse theory. Dorrance Publishing Co., Inc., Pittsburgh, USA. 118p.

Simpson, J.F., 1967. Solar activity as a triggering mechanism for earthquakes. *Earth and Planetary Science Letters*, v. 3, p. 417–425.

Tsunoda, F., 2010. Habits of earthquakes. Part 2. Earthquake corridors in East Asia. *NCGT Newsletter*, no. 54, p. 45–56.

Zhang, G-Q., 1998. Relationship between global seismicity and solar activities. *ActaSeismologicaSinica*, v. 11, p. 495–500.

Appendix 2, Item 3

Earthquake/Volcanic Activities and Solar Cycles

A commentary by Dr. Dong Choi
Director of Research, International Earthquake and Volcano Prediction Center (IEVPC), dchoi@ievpc.org
Editor-In Chief, New Concepts in Global Tectonics Journal (NCGT)

Abstract

Recent studies have established that Earth's geodynamic events such as earthquakes and volcanic eruptions are affected chiefly by solar cycles. An anti-correlation is seen between the solar activity cycle and earthquake frequency. Many of the catastrophic earthquakes and volcanic eruptions have occurred during the major solar low or lowering periods—Maunder Minimum (1645-1715), Dalton Minimum (1793–1830), "1900 trough" (1880–1915, *new name*), and the declining period of the long-term solar cycles, 1990–2013.

Furthermore, another yet-unknown planetary force overrides the solar influence, as seen in a conspicuous seismo-volcanic quiescence from 1996 to 2003. These facts imply that the Earth's tectonic events and hence climate are strongly affected by the Sun and other planetary systems, and that the Sun is the most powerful global climate driver. Long-term solar cycles must be taken into account to predict future climate trends.

Keywords: *earthquake, volcano, solar cycle, Maunder and Dalton Minimums, seismic-volcanic quiescence*

1. Introduction.

In recent years, researchers have clarified the intricate interaction between solar activity and Earth's geodynamic events (earthquakes and volcanic eruptions). What has become clear is the reversed correlation between solar and earthquake cycles, and heightened seismic and volcanic activities during major solar low cycles—Maunder (1645–1715) and Dalton (1793–1830) Minimums or Little Ice Ages (Casey, 2010 and 2012; Choi, 2010; Choi and Maslov, 2010; Choi and Tsunoda, 2011).

Casey (2010 and 2012) found 87.5% of large-scale global volcanic eruptions (Volcanic Eruption Index, or VEI, 5 or greater), took place during major solar minimums. He also noted strongest earthquakes (M7.8 or greater) in the continental USA occurred during major solar activity minimums. Other earlier studies by Stothers (1989) also found that incidence of volcanic eruptions is slightly greater around the time of solar minimum than at any other phase. Both Stothers (1989) and Fairbridge (1980) noted the abnormally high volcanic eruption numbers during the Maunder Minimum. This is supported by elevated acidity in Greenland deep ice cores which cover the years from 533 to 1972 AD during this protracted solar minimum (Hammer et al., 1980).

As stated above, the heightened volcanic and seismic activities during solar low periods are almost beyond reasonable doubt. In this short article the author reviews the current understanding on this subject with some additional new information.

2. Solar and earthquake/volcanic cycles.

1) Anti-correlation between the solar and earthquake cycles.

Earthquake data retrieved from the USGS NEIC archives (http://earthquake.usgs.gov/earthquakes/eqarchives/epic/), were studied extensively by the author (Choi, 2010; Choi and Maslov, 2010). A comparison of the earthquake frequency curve with solar cycles revealed a very interesting trend (Fig. 1)—an anti-correlation. This means that when solar cycle is high the number of earthquakes drops, whereas when solar activity is low, more earthquakes tend to occur. This reversed correlation has been validated by many researchers afterwards (Tsunoda et al. 2013 for example).

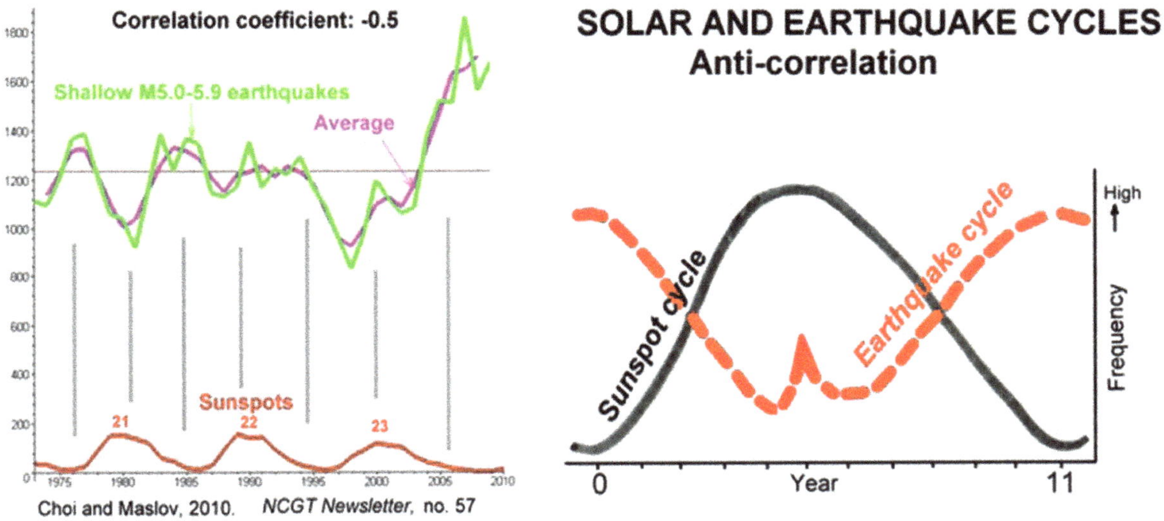

Figure 1. Anti-correlation between solar and earthquake cycles. A group of shallow (100 km or shallower) earthquakes with magnitude 5.0–5.9 was used for analysis. Deeper and longer trough during the solar cycle 23 is due to seismic dormancy possibly caused by other planetary forces as described in the text.

2) Solar and volcanic cycles.

Volcanic activities, on the other hand, show two contradictory trends (Fig. 2; Tsunoda et al., 2013); a positive correlation with the solar curve for worldwide volcanoes, whereas a reversed correlation for volcanoes in Southwest Pacific where massive energy rises directly from the outer core through asuperplume. Historically known strongest volcanic eruptions, VEI 5 or greater, also show similar trend as those in SW Pacific—anti-correlation with the solar cycle (Figs. 3 & 4).

The cause of this anti-correlation awaits further study. One of the most feasible explanations was presented by Gregori (2002) who attributed to the Earth's core being a leaky capacitor or a battery; when solar activity is high, the Earth's core is charged, whereas when the Sun's activity is in low phase, the core in turn discharges energy. Another theory is cosmic rays; when the solar activity is low, the amount of ionized cosmic rays with stronger penetration capability increases (Kirby, 2007, and others); the increased cosmic rays may heat up the Earth's interior to discharge more energy.

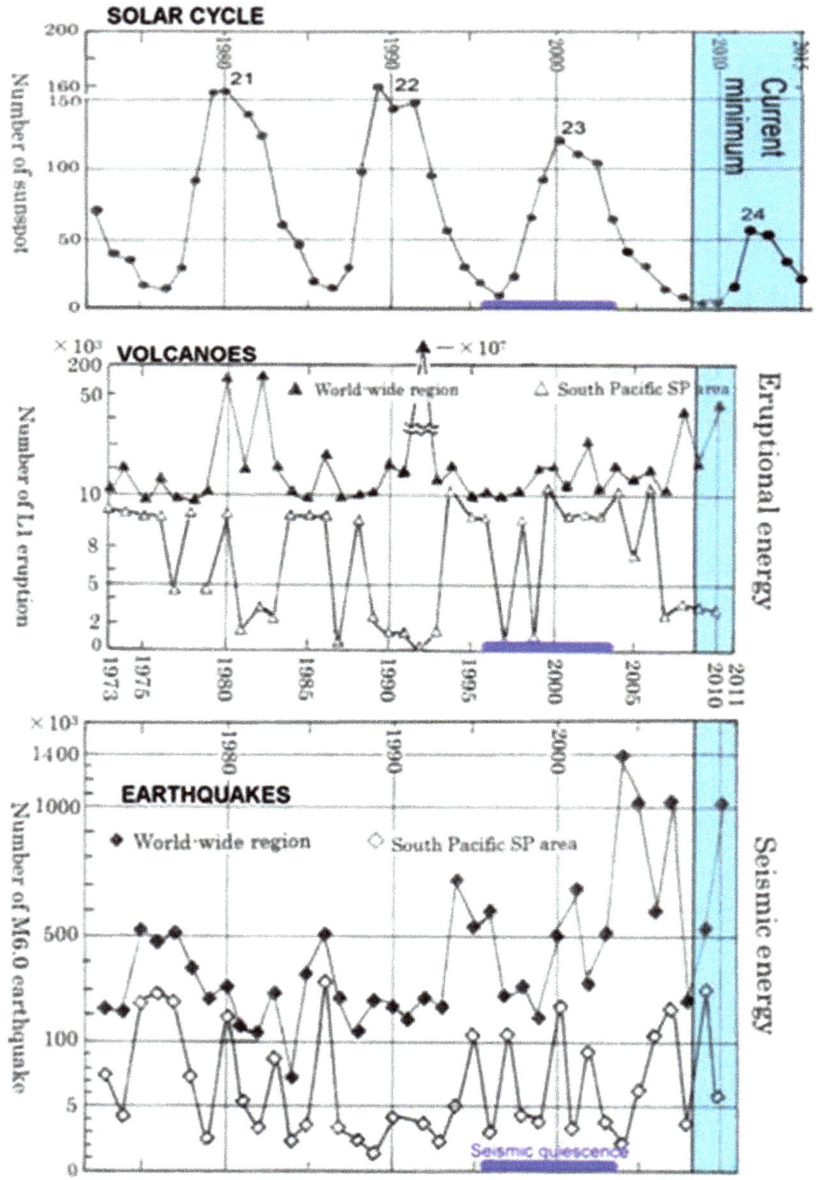

Figure 2. Earthquake and volcanic energy variations from 1973 to 2010. Note contradictory volcanic energy fluctuation pattern between the worldwide and SW Pacific regions in the middle figure. Seismic quiescence (1996–2003, Choi, 2010) is shown by a blue bar. The worldwide volcanic fluctuation trend (filled triangle in the middle figure) is also low during the same period. Earthquakes in both worldwide and SW Pacific Superplume (SP) regions show similar pattern and a reversed correlation with solar cycles is recognized. Sunspot numbers after 2012 taken from Solar Influences Data Analysis Center (http://sidc.oma.be). Cited from Tsunoda et al., 2013.

3) Historic strong seismic and volcanic activities in relation to solar cycle since 1600 to present

Our studies (Choi and Tsunoda, 2011; Choi, in preparation, 2003) clarified that a massive number of very strong earthquakes and volcanic eruptions occurred during the Maunder Minimum (Fig. 3). The seismic and magmatic events dwindled dramatically after the end of the Maunder Minimum. The Dalton Minimum shows some regional variations; seismically still quiet in Japan and Turkey, whereas in some other regions including India, Indonesia, continental North America and northern South America very strong earthquakes occurred. The strongest (VEI 7) Indonesian Tambora volcano erupted in this solar low period, or Dalton Minimum. Magmatic and seismic events again became active worldwide after the closure of the Dalton Minimum; this trend has continued until today with a conspicuous rise since 1990 when deep strong earthquakes have started to appear—the year coincides with the start of sharp decline of the 206year solar cycle (Casey, 2010).

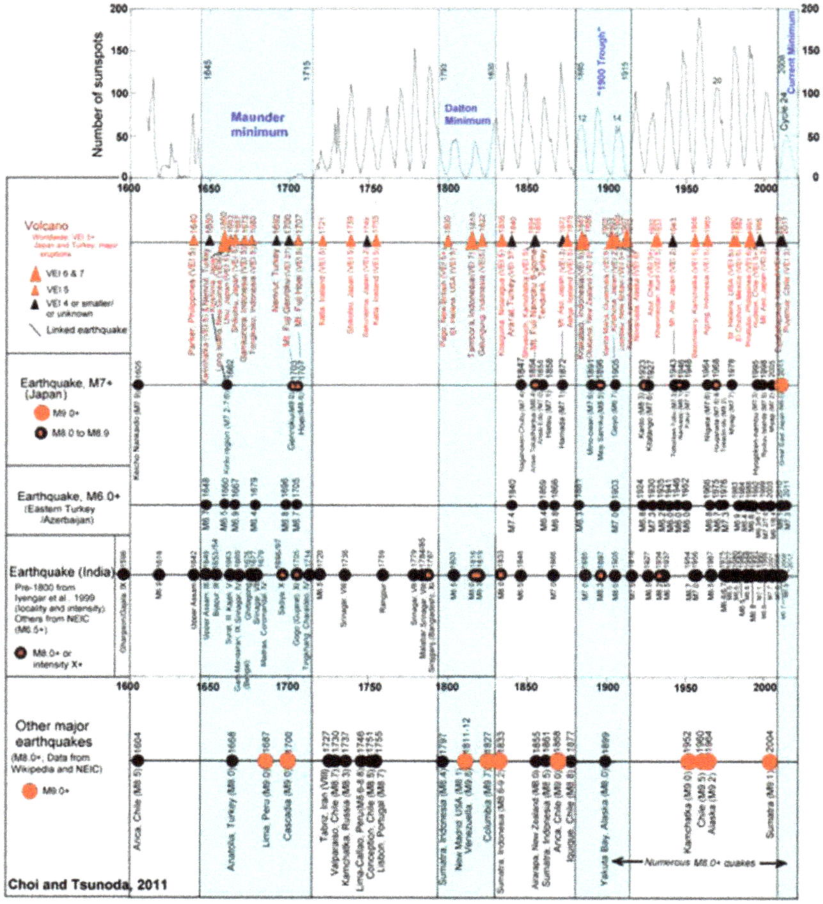

Figure 3. World major volcanic eruptions and earthquakes in comparison with solar cycles. Cited from Choi and Tsunoda (2011).

Another more systematic analysis of volcanic records was conducted for major volcanoes with extensive historic documentation in all Circum-Pacific countries (Fig. 4). A total of 195 volcanoes were selected, and each Volcanic Explosive Level (VEI) was converted to numbers to facilitate quantitative treatment. A unique method was adopted for this quantification as used for the study of historic Indonesian volcanoes, see Choi and Tsunoda (2011).

As clearly seen in the figure, majority of large eruptions with VEI 5 or greater are concentrated in the major solar lows or rapidly declining periods of the bi-centennial or centennial solar cycles, 206 and 90–100 year cycles, especially from 1990 to 2012. Even those with VEI 5+ eruptions which do not fall in these low or lowering cycle windows, if seen carefully, however, occurred mostly in the trough of a short cycle (11 year—Schwabe cycle).

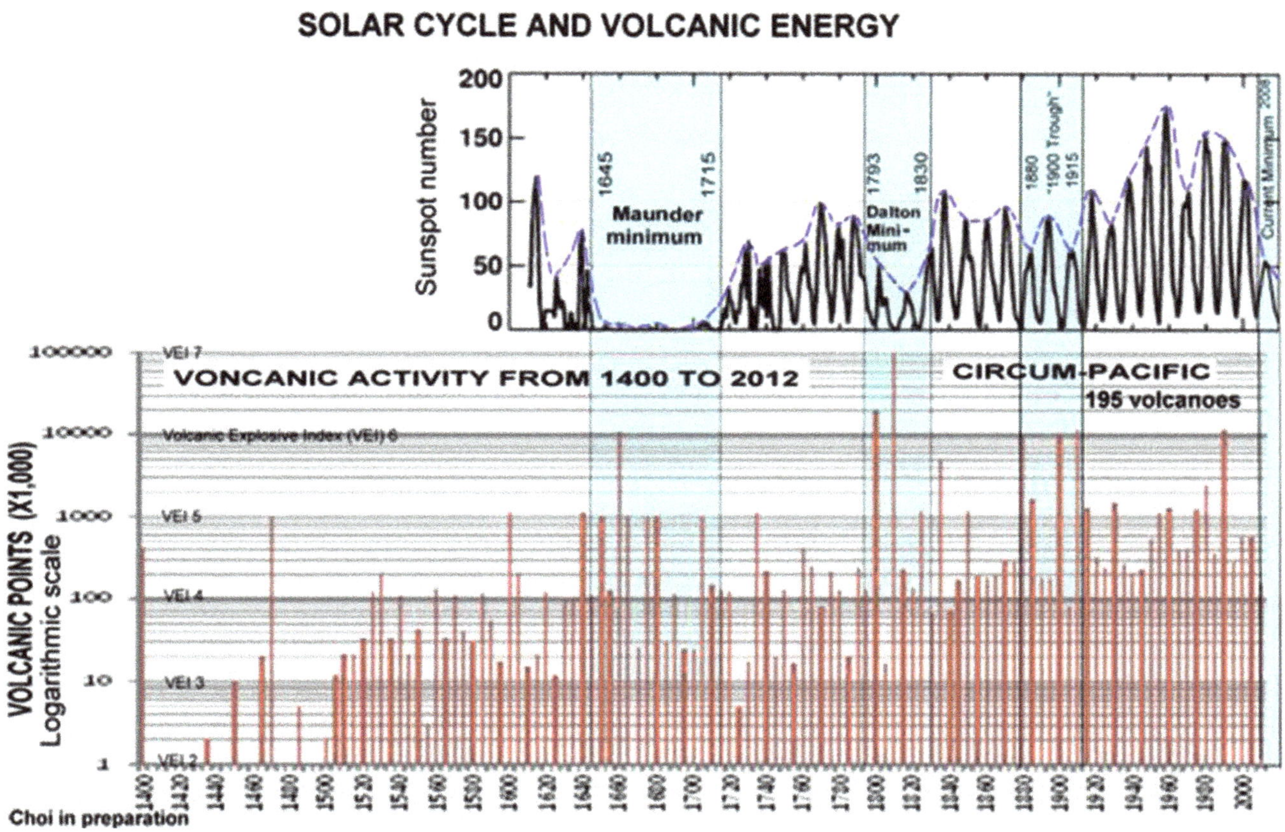

Figure 4. Fluctuation of historic strong volcanic activity in the Circum-Pacific regions in comparison with solar cycles. Volcano data from Smithsonian Institution's Global Volcanism Program (http://www.volcano.si.edu/index.cfm). Each red bar represents the total volcanic energy level in a five-year window.

3. Influence of another planetary force on volcanic and earthquake activities.

In addition to the above solar influence on Earth's geodynamic events, another planetary force is obviously affecting earthquake and volcanic activities. This was concluded based on a clear seismic quiescence period from 1996 to 2003 independent of solar cycle by the present author (Choi, 2010; Choi and Maslov, 2010). As seen in Figure 5, the occurrence of worldwide strong earthquakes with magnitude 5.0 or greater suddenly dropped by up to 40% at the beginning of 1996. The dormant period lasted until the end of 2003, or a span of eight years before the dramatic rise in the early 2007; it looks as if one of the Earth's switches went off during the period. This unusual period covers the trough and first two-thirds of solar cycle 23, which does not match the anti-correlation pattern between the 11-year solar cycle and earthquake frequency. If it was in normal cycle, earthquake frequency would have continued to rise to 1996 when the solar cycle bottomed, and to decrease gradually after that.

This large seismically subdued period is also observed in the worldwide volcanic energy level fluctuation from 1995 to 2007 (filled triangles, middle figure of Fig. 2)—for a duration of 12 years. This prolonged quiet period covers from the trough between solar cycles 22 and 23 to the lowering time of cycle 23—entire solar cycle 23. Interestingly prior to 1995 or during solar cycles 21 and 22, solar and world volcanic cycles (excluding the SW Pacific superplumearea) had been positively correlated.

As Choi (2010) observed in the world earthquake trend, the quiescence is recognized in all regions except for Chile, and it is observable mainly for shallow quakes with magnitude 6.0 or smaller. The quiescence is most dramatically observed in the SW Pacific in relatively small quakes below M6.0 but it affects all depths in this particular region (Fig. 5 top figure). Because the SW Pacific is the area where the superplumes rise directly from the outer core (Tsunoda et al., 2013), it may imply that the effect of the external force has penetrated deeper into the Earth in this particular area. In this connection, it should be noted that volcanoes in this particular area show anti-correlation with solar cycle—same as earthquake cycles (see Fig. 2). Also noteworthy is that almost all of the gigantic volcanic eruptions took place during the solar low periods when the Earth's core discharged strong energy.

The cause of this seismic/magmatic quiescence along with that of the Sun-earthquake anti-correlation must be investigated by a multi-disciplinary team. This is a fascinating field of study, and its results will give a better understanding of the Earth's geodynamic processes and their interaction with the Sun and other planetary systems.

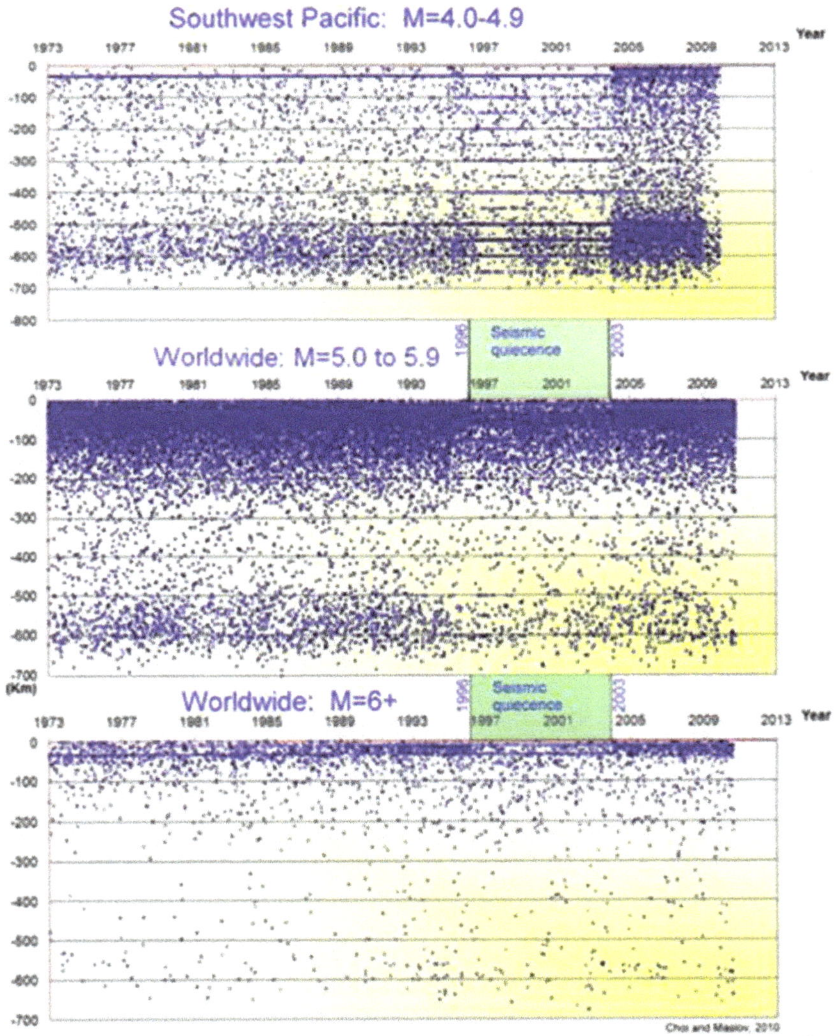

Figure 5. Time-depth plot of earthquakes from 1973 to 2010 showing the seismic quiescence. In all magnitudes and depths of quakes below M7.0 worldwide, this quiescent period is observed, especially most distinctively in quakes smaller than M6.0 in the SW Pacific superplume region. However, deep strong quakes are relatively little affected by this quiescence, implying the quiescence was caused by an external force, but not coming from the Earth's core. If the core of the Earth was the cause of the quiescence, the effect must be felt at all depths evenly. For more details, see Choi, 2010 and Choi and Maslov, 2011.

4. Conclusion

This paper introduced how the Sun and other powerful planetary forces are intricately interacting with the Earth's geodynamic events, represented by earthquakes and volcanic eruptions. Long-term solar cycle trend is most crucial in understanding the past global climate, and hence in predicting future climate trend. As advocated by Casey (2012 and 2013a & b), it is doubtless that the Earth has entered a major low cycle or "hibernation" era, and we will have more catastrophic earthquakes and volcanic eruptions in the coming several decades.

References

Casey, J., 2010. Correlation of solar activity minimums and large magnitude geophysical events. *Space and Science Research Center, Research Report* 1-2010 (Preliminary), p. 1–5.

Casey, J.L., 2012. *Cold sun*. Trafford Publishing, 167p.

Casey, J.L., 2013a. Global climate status report (GCSR). Edition 1-2013. 51p. Space and Science Research Corporation (SSRC). www.spaceandscience.net.

Casey, J.L., 2013b. Global climate status report (GCSR). Edition 2-2013, 67p. Space and Science Research Corporation (SSRC). www.spaceandscience.net.

Choi, D.R., 2010. Global seismic synchronicity. *NCGT Newsletter*, no. 55, p. 66–73.

Choi, D.R. and Maslov, L., 2010. Earthquakes and solar activity cycles. *NCGT Newsletter*, no. 57, p. 85–97.

Choi, D.R. and Tsunoda, F., 2011. Volcanic and seismic activities during the solar hibernation periods. *NCGT Newsletter*, no. 61, p. 78-87.

Gregori, G.P., 2002. Galaxy-Sun-Earth relations. The origin of the magnetic field and of the endogenous energy of the Earth, with implication for volcanism, geodynamic and climate control, and related items of concern for starts, planets, satellite, and other planetary objects. A discussion in a prologue and two parts. BeiträgezurGeoschichte der Geophysik und KosmischenPhysik, Band 3, Helt 3, 471p.

Fairbridge, R.W., 1980. Prediction of long-term geologic and climatic changes that might affect the isolation of radioactive waste. In, "Underground disposal of radioactive wastes", v. 2, p. 385–405. International Atomic Agency, Vienna.

Hammer, C.U., Clausen, H.B. and Dansgaard, W., 1980. Greenland ice sheet evidence of post-glacial volcanism and its climatic impact. *Nature*, v. 288, p. 230–235.

Iyengar, R.N., Sharma, D. and Siddiqui, J.M., 1999. Earthquake history of India in medieval times. *Indian Journal of History of Science*, v. 34, no. 3, p. 181–237.

Kirby, J. 2007. Cosmic rays and climate. *Surveys in Geophysics*, v. 28, p. 333–375.

Stothers, R.B., 1989. Volcanic eruptions and solar activity.*Jour. Geophys. Res.* V. 94, no. B12, p. 17,371–17,381.

Tsunoda, F., Choi, D.R. and Kawabe, T., 2013. Thermal energy transmigration and fluctuation. *NCGT Journal*, v. 1, no. 2, p. 65–80.

Appendix 2, Item 4

Earthquakes and Solar Cycles: Increased Earth Core Activity Since 1990

A commentary on major research findings that show a strong correlation between solar activity and geophysical events like earthquakes
By Dr. Dong Choi,[1] Mr. John Casey,[2] Dr.Leo Maslov,[3] and Dr. Fumio Tsunoda[4]

This short summary introduces a glimpse of some significant findings from ongoing studies focusing on the past 44 years and the relationship of solar activity to earthquakes/volcanic eruptions. A full paper along these lines will be published in the future. **Importantly, the authors have established that solar activity and geophysical activity are strongly correlated.**

It was determined during the combined research of the authors that:

1. **There exists a strong, yet inversed correlation between the 11-year solar cycle (Schwabe Cycle) and the total number of earthquakes**. When the Sun is in active phase, earthquakes occur less frequently, but when in low phase more earthquakes occur (Choi, 2010; Choi and Maslov, 2010; Choi, 2013; Tsunoda et al., 2013). The follow-up studies by us and others have confirmed this general trend.

---

[1] . International Earthquake and Volcano Prediction Center (IEVPC), Australia. dchoi@ievpc.org
[2] . Space and Science Research Corporation/IEVPC, USA. jcasey@ievpc.org
[3] . Department of Physics and Astronomy, University of Northern Colorado, USA. lev.maslov@unco.edu
[4] . Professor (emeritus), Saitama University, Japan. tsunochan@sky.email.ne.jp

2. **We have determined that there are clear trends in the Earth's geodynamic cycle represented by an overall active Earth core phase from 1990 onwards.** This is demonstrated in several subordinate trends:

a. a. Pronounced increase in deep quakes of M6.0+ (depth =300km +)

b. b. Increased shallow intermediate quakes of M6.0+ (depth = 0-300km)

c. c. A representative trend of increased seismic activity in California since 1990.

3. A negative correlation has been found between core activity and solar cycles over the 44 year period of observation: This is observed in:

a. An increase in geophysical activity with declining solar activity over the selected 44 year period.
b. A unique period of seismic/volcanic quiescence was seen during solar cycle 23.

4. Unique relationships were detected between the solar cycles and their earthquakes and volcanic eruptions during the period of observation:

a. Though major volcanic eruptions (VEI 5+) were observed during the peaks of solar cycles 21, 22, they were absent during solar cycles 23 and 24. The latter two cycles represent the declining phase of solar activity for the last major solar cycle group (cycles 14-24). Solar cycle 14 was at peak, though a relatively low one in February 1906 with a sunspot count of about 64. This corresponds to a similar sunspot low during cycle 24.

b. Historically large earthquakes that occurred in Sumatra (2004) and Japan (2011) and may represent confirmation of concurrent major earthquakes with centennial solar activity reductions as previously identified by the authors.

The increasing seismic activity since 1990 is expected to continue for the coming two to three decades as we have entered a "solar hibernation" or possibly a mini-ice age (Casey, 2012); this will likely bring more strong, possibly catastrophic earthquakes.

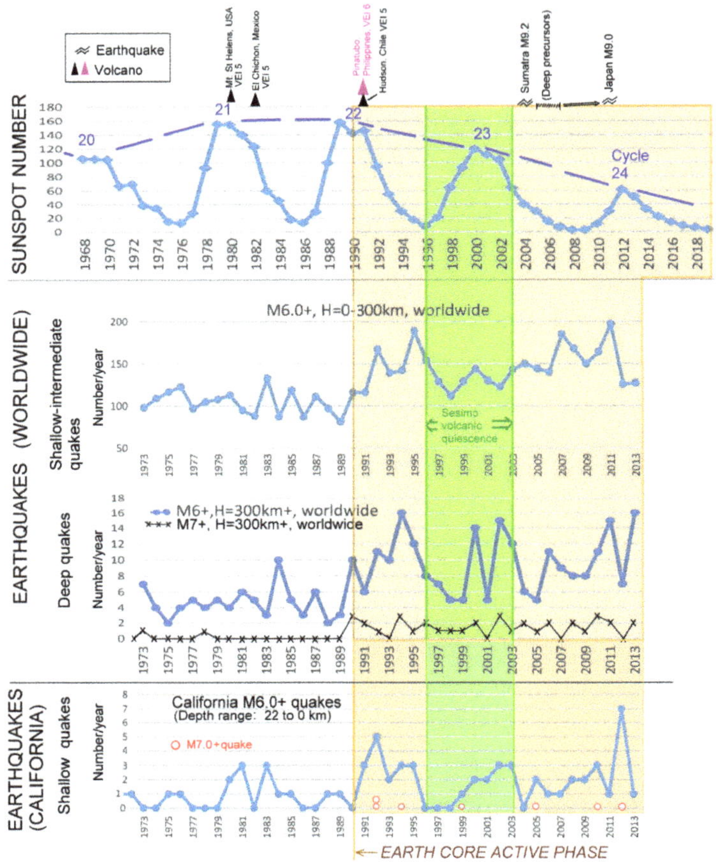

Figure 1. Solar cycles and earthquakes.

The correlation coefficient between the worldwide earthquakes and the 11-year solar cycle is -0.5, a significant negative correlation. Note an outstanding contrast in the worldwide earthquake trend before and after 1990. California quakes also show the same trend—represented by an abrupt appearance of M7.0+ earthquakes and a sudden increase in M6.0+ quakes. Note an overall rise in earthquake number during the 44 year period. This trend is expected to continue. Seismo-volcanic quiescence cited from Choi (2010) and Tsunoda et al. (2013), and the Earth core active phase from Choi and Maslov (2010). Sunspot numbers after 2012 taken from Solar Influences Data Analysis Center (http://sidc.oma.be). Earthquake data from IRIS and USGS NEIC.

References

Casey, J., 2010. Correlation of solar activity minimums and large magnitude geophysical events. *Space and Science Research Center, Research Center, Research Report* 1- 2010 (Preliminary), p. 1–5.

Casey, J., 2012. Cold Sun. Trafford Publishing, 167p.

Choi, D.R., 2010a. Global seismic synchronicity. *NCGT Newsletter,* no. 55, p. 66–73.

Choi, D.R., 2013. Earthquake/volcanic activities and solar cycles. *Global Climate Status Report (GCSR)*, Space and Science Research Corporation, Edition 3- 2013, p. 10–19.

Choi, D.R. and Maslov, L., 2010. Earthquakes and solar activity cycles. *NCGT Newsletter,* no. 57, p. 85–97.

Quinn, J.M., 2010. Global warming. Geophysical counterpoints to the enhanced greenhouse theory. Dorance Publishing Co., Inc., Pittsburgh, USA. 118p.

Tsunoda, F., Choi, D.R. and Kawabe, T., 2013. Thermal energy transmigration and fluctuation. *NCGT Journal,* v. 1, no. 2, p. 65–80.

Appendix 2, Item 5

The New Madrid Seismic Zone, Central USA: The Great 1811–1812 Earthquakes, Their Relationship to Solar Cycles and Tectonic Settings

Dong R. CHOI
Raax Australia Pty Ltd. Dong.Choi@raax.com.au; www.raax.com.au
International Earthquake and Volcano Prediction Center. dchoi@ievpc.org; www.ievpc.org

John L. CASEY
Space and Science Research Corporation, mail@spaceandscience.net; www.spaceandscience.net
International Earthquake and Volcano Prediction Center,jcasey@ievpc.org; www.ievpc.org

Abstract

The 1811–1812 New Madrid series of earthquakes were the largest in magnitude (estimated to be M8.0 or greater) in the continental North America in the history. The quakes occurred in the midst of Dalton Solar Minimum (1793–1830). Other major historic earthquakes in the same region also occurred during major solar minimums, or "solar hibernations." From a tectonic viewpoint, the New Madrid Seismic Zone (NMSZ) is situated on the axis of the N-S American Geanticline or Super Anticline which is Archean in origin. It has been subject to repeated magmatic and tectonic activities in Proterozoic and Phanerozoic—the Caribbean dome (now oceanized to form the Caribbean Sea and the Gulf of Mexico) has been the site for rising thermal energy from the outer core since the Mesozoic. Energy transmigrates northward along the anticlinal axis (or surge channel) and is trapped at the embayment bounded by less permeable Precambrian-Paleozoic basement highs in the north of the New Madrid area. The arrival of a major, prolonged solar low period or "hibernation" in the coming 30 years, which are considered comparable to the Dalton or even Maunder Minimum (1645-1715), increases the likelihood of repeating the 1811–12 class seismic events. Heightened awareness, monitoring of precursory signals, and disaster mitigation planning are required.

Keywords: *1811–12 New Madrid Earthquakes, Dalton Minimum, solar hibernation, N-S American Super Anticline, surge channel, seismic energy transmigration, earthquake-solar cycle anti-correlation*

Introduction

The New Madrid area, mid-Mississippi River, central United State, was rocked by a spate of powerful earthquakes from 1811 to 1812 (**Fig. 1**). According to the USGS records, there were three main shocks, M7.5, 7.3 and 7.5, on 16 December 1811, 23 January 1812, and 7 February 1812, respectively, with a major aftershock M7.0 on the first day (http://earthquake.usgs.gov/earthquakes/states/events/1811-1812.php). Other researchers, such as Nuttli (1987) listed six M7.0+ quakes that include two M8.0+ earthquakes. Of them, two largest quakes were considered the greatest earthquakes in continental North America (Johnston and Schweig, 1996).

The sequence of the great earthquakes in the NMSZ has a unique attribute—it occurred in the middle of a major solar low period, Dalton Minimum, 1793 to 1830 (**Fig. 2**). This prompted the authors to study seismic history of the NMSZ and their relation to solar cycles, together with geological settings of the surrounding region. The rationales of this study are, 1) the arrival of a prolonged solar low period as advocated by Casey (2008, 2012 and 2014), and 2) the well-established reversed correlation between the solar activity cycle and earthquake energy (Choi and Maslov, 2010), and 3) new interpretation of geological structure of the region and seismic energy transmigration mechanism in the Caribbean-Gulf Mexico-Mississippi River (Choi, 2013; Choi, 2014; Choi et al., 2014).

Seismic activity in the NMSZ and solar cycles

Historic records show that the New Madrid region has been subject to repeated seismic activities. Based on artefacts found buried by sand blow deposits and from carbon-14 studies, previous large earthquakes like those of 1811–1812 appear to have happened around 4800BC, 3500BC, 2350 BC, AD300, AD900 and AD1450. In addition, the first known written record of an earthquake felt in the New Madrid Seismic Zone occurred on Christmas Day of 1699. An M6.6 earthquake in 1895 has also been registered (Wikipedia, http://en.wikipedia.org/wiki/New_Madrid_Seismic_Zone).

Most of the years listed above belong to solar low periods (**Figs. 2 and 3**): The years 1811–1812 is in the midst of a major solar low period, Dalton Minimum. The year 1699 sits in another major solar low period, Maunder Minimum, 1645–1715. AD1450 corresponds to the lowering period of Spörer Minimum, and another one in 1895, centennial low cycle (1885–1915; Casey, 2008; **Fig. 2**).

Importantly, all major Earthquakes in the NMSZ since 1400 AD have occurred during these solar low points or solar hibernations.

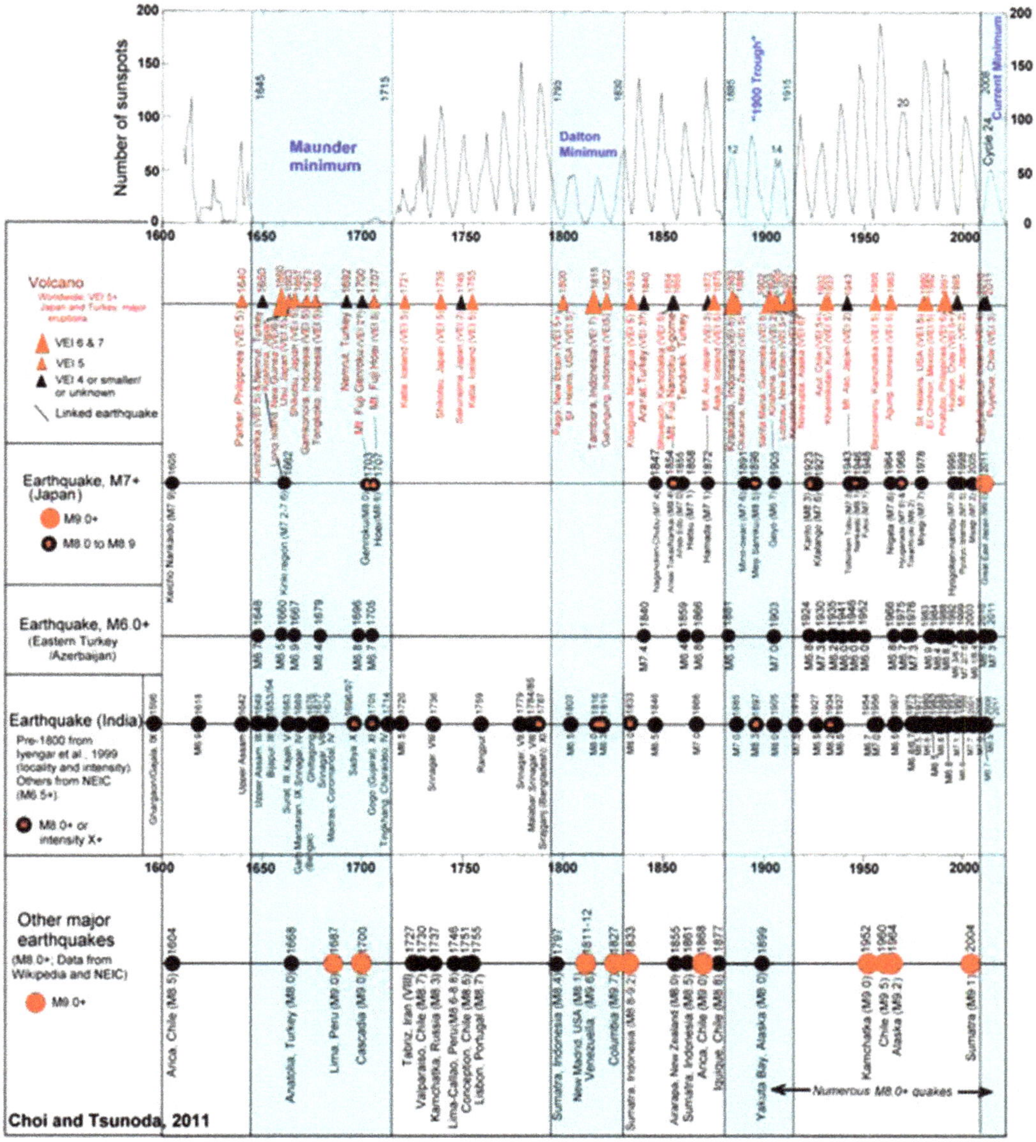

Fig. 2. Solar cycle and world volcanic/seismic activities. All of the NMSZ quakes occurred around the middle of the solar low periods. Cited from Choi and Tsunoda, 2011 and Choi, 2013b.

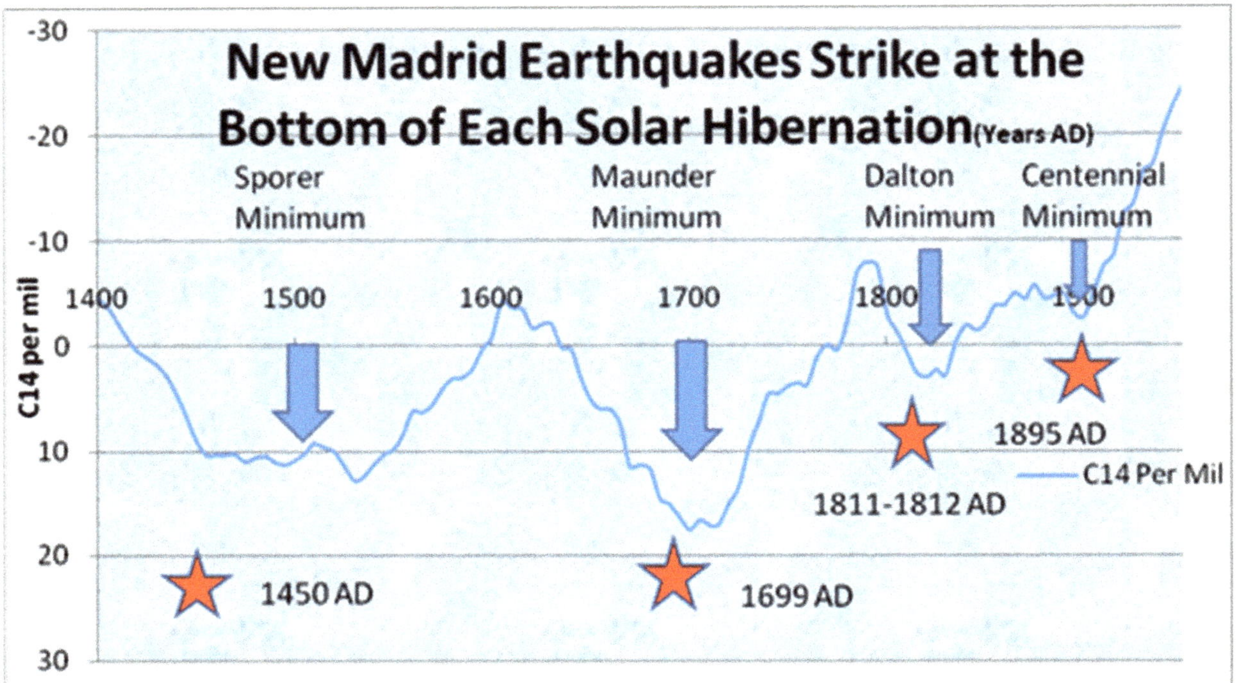

Fig. 3. History of New Madrid earthquakes compared to solar minimums or "solar hibernations" from 1400–1950 AD. Solar activity deduced from C[14]proxy variation.The years of major New Madrid earthquakes are shown in red stars with dates. Source: Casey, Data: Reimar et al. INTCAL04

The NMSZ quakes and solar cycles indicate their reversed correlation. The anti-correlation between solar cycles and seismic/volcanic activities has been well established by the senior author of this paper with co-workers (**Fig. 4**; Choi and Maslov, 2010; Choi and Tsunoda, 2011). Casey (2010) also noted that the catastrophic volcanic eruptions had taken place during the solar low periods.

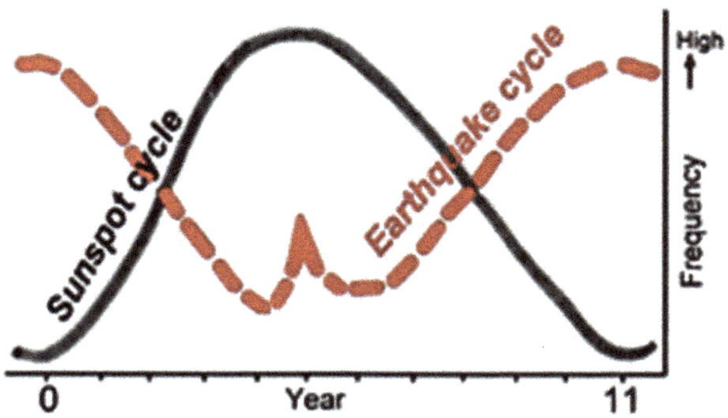

Fig. 4. Anti-correlation between the solar and earthquake cycles (Choi and Maslov, 2010).

The cause of this anti-correlation awaits further study. One of the feasible explanations was presented by Gregori (2002) who attributed to the Earth's core being a leaky capacitor or a battery; when solar activity is high, the Earth's core is charged, whereas when the Sun's activity is in low phase, the core in turn discharges energy.

Discussion

1) Geological structures responsible for the NMSZ earthquakes.

The earthquakes occurred in the NMSZ come from the unique tectonic settings. It is strongly related to the global-scale geological structure; North-South American Geanticline or Super Anticline that runs from South America, via the Caribbean and Mississippi Valley, to the Canadian Shield (Choi, 2013; **Figs. 5 and 6**). It is a fundamental geological structure formed in the early stage of the Earth's formation—in Archean. There is another antipodal super anticline that extends from SW Pacific, via SE Asia and South China, to Siberia. These anticlinal structures have influenced the subsequent development of the Earth by repeated magmatic and tectonic activities throughout the Phanerozoic, especially since Mesozoic.

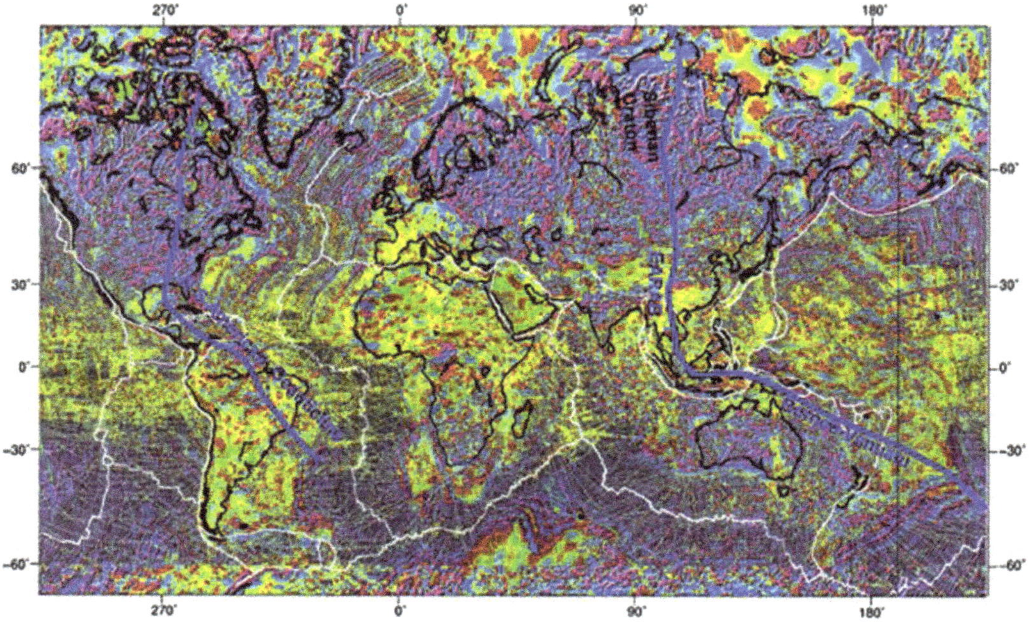

Fig. 5. Earth's fundamental structures; two antipodal super anticlines (Choi, 2013a). Note that the Caribbean Sea and the Mississippi Valley are situated on the axis of the anticline. Base map, World magnetic anomaly map, by Korhonen et al., 2007.

In his 2010 and 2014 papers, the senior author argued the origin of the Caribbean-Gulf of Mexico, which developed in the axial part of the anticline and formed the Caribbean dome; the crust in the site where energy rose from the outer core has been oceanized since Mesozoic. The initial basin formation however may go back to Paleozoic time (Pratch, 2008 and 2010). The axial area, being highly fractured and permeable, became a channel of energy flow, or surge channel (Meyerhoff et al., 1996). The thermal seismic energy, derived from the outer core through the Caribbean dome and transmigrated along the surge channel developed under the Mississippi Valley, is responsible for the NMSZ earthquakes (**Fig. 6**). This assertion is supported by the fact that, along the Pacific coast of Central America, the seismo-volcanic energy which was originated from the deep Caribbean was found to transmigrate northward during the solar low cycles but southward during the rising cycles (Choi, 2014). The energy from the outer core was stronger during the time of solar low phase, as evidenced by the well-established solar cycle-earthquake anti-correlation (**Fig. 4**).

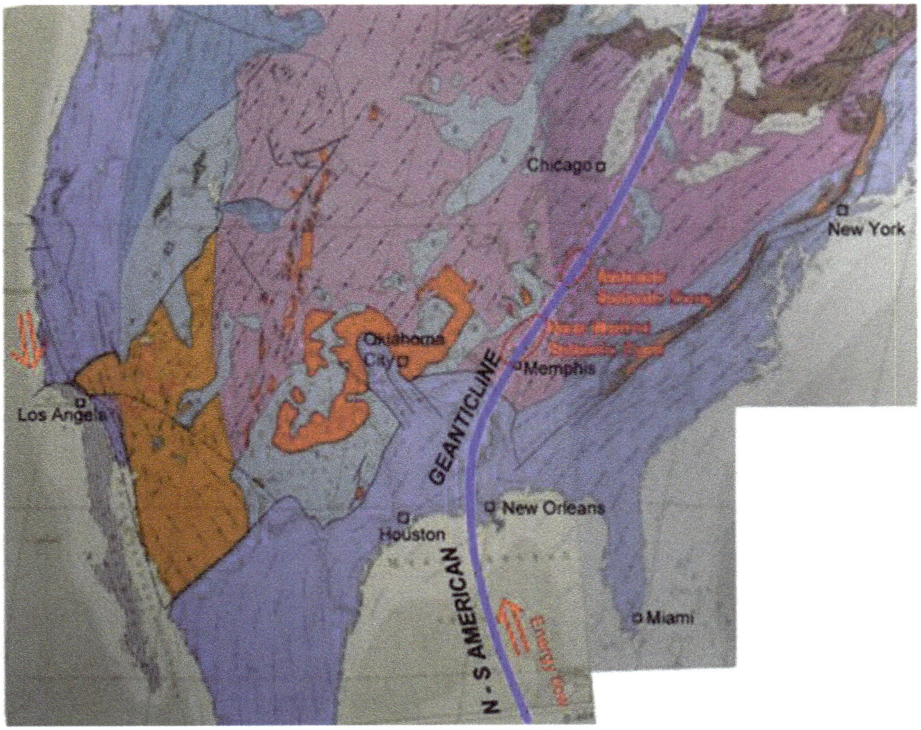

Fig. 6. N-S American Geanticline, the NMSZ and deep structure of the North America represented by Precambrian structures (Kosygin et al., 1970). Energy flow direction along the N-S American Geanticlinal axis from Choi (1014), and for California-Mexico from Choi et al. (2014). Note the prevailing NE-SW deep structural trends which seemingly continue into the Pacific Ocean.

A geological map, **Fig. 7**, well illustrates a Mesozoic embayment developed along the Mississippi Valley. The NMSZ area is the northern end of the Mesozoic basin that covers the present Gulf of Mexico and the Caribbean. The NMSZ region is surrounded by older, less permeable, Precambrian-Paleozoic rocks—which form a trap structure for thermal seismic energy in the form of liquid and gas. The trap structures were controlled by deep fault systems, which are NE-SW and NW-SE in direction (Johnson and Schweig, 1996).

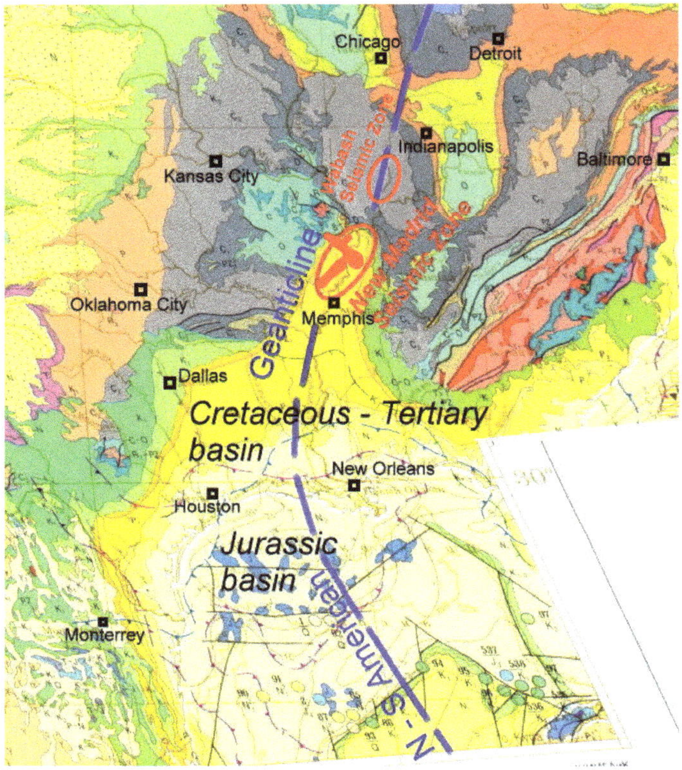

Fig. 7. Geologic map by Jatskevich et al. (2000) superimposed by tectonic elements and the NMSZ which is located at the northern end of the Mesozoic-Paleogene basin (labelled as K, K_1, K_2 and ₽).

2) Arrival of a major, prolonged, solar low period, or solar hibernation.

The correlation of major earthquakes and solar activity, while relatively recently discussed, is nonetheless one of the strongest in terms of climate change and geophysical associations. The initial paper (Casey 2008) on the regular pattern of climate oscillations linked to solar activity using the Relational Cycle Theory (RC Theory) has demonstrated itself to be among the most successful in climate prediction underscoring the basic reliability of the theory and its associated seven elements of climate change. Subsequently (Casey 2010) in a preliminary paper, proposed the connection between the RC Theory and major earthquakes and volcanic activity. Others noted above (Choi, Maslov, et al.), have also found the strong relationship between solar activity lows and increased seismic and volcanic activity.

Conclusions

This study revealed several important factual data regarding the strong earthquakes in the NMSZ and their relation to solar cycle. It also presented new interpretation of tectonic settings of the region. They are summarized as follows:

1. The NMSZ developed on the major Precambrian-origin geanticlinal axis where magmatic, thermal, and tectonic activities have been concentrated, particularly since Mesozoic when the Gulf of Mexico and the Caribbean have started to form. This activity is still continuing today.

2. The historic record clearly shows that large seismic events in the NMSZ have occurred during the Sun's inactive periods. The sequence of 1811–12 quakes is one of them.

3. In the light of the now confirmed start of a prolonged, solar hibernation for the coming 30 years or so, which are comparable to Dalton Minimum or worst case, a Maunder Minimum ("Little Ice Age"), a repeat of the 1811–12 earthquakes should be expected.

4. The window of highest risk for another major New Madrid zone earthquake is between 2017 and 2038.

5. Planning for a repeat of the 1811–1812 series of earthquakes that devastated the region back then should begin immediately. Considerations should include:

a. A US nationwide plan is required based on one or more M8.0+ earthquakes in the NMSZ on the assumption that substantial regional loss of life and massive infrastructure damage will take place on a scale never before witnessed in the USA.

b. This plan should include heightened levels of public education, monitoring of the seismic precursory signals, federal, state and local emergency management exercises and damage mitigation where practicable.

c. Planning should address the real possibility of complete loss of major ground and air transportation nodes and routes including substantial long term damage to airport facilities and runways and interstate and city highway systems especially across the Mississippi River.

d. Planning should also include the assumption that major aftershocks will prevent meaningful rebuilding of permanent structures over several months to a year.

e. Should a repeat of a series of quakes take place similar to the 1811–1812 events or even a repeat of the 1895 M6.6 earthquake, the power grid in the central Mississippi region may be unavailable for essential needs of radio and TV communications, emergency management, search and rescue etc. for several months to a half year or more.

f. In the case where there may be NMSZ nuclear facilities not designed to withstand a series of M7.5 to M8.0+ earthquakes, a new added risk may exist. All nuclear facilities must be reviewed (if not already done so) to insure they and their back-up power systems for coolant systems etc., can withstand a worst case series of major quakes. Failure to do so could result in multiple instances of the March 11, 2011 Japanese, Fukushima nuclear reactor style catastrophes in the middle of the United States. This could directly affect the safety of all citizens east of the central Mississippi River subject to prevailing winds during the time of the year such a scenario might happen.

References cited

Cahill, R. T., 2014. Solar flare five-day predictions from quantum detectors of dynamical space fractal flow turbulence: Gravitational wave diminution and Earth climate cooling. *Progress in Physics*, v. 10, Issue 4 (October), p. 236–242. (http://ptep-online.com/index_files/2014/PP-39-10.PDF).

Casey, J. L., 2008. The existence of 'relational cycles' of solar activity on a multi-decadal to centennial scale, as significant models of climate change on Earth. *Space and Science Research Center, Research Report 1-2008* – The RC Theory. p. 1-8. www.spaceandscience.net

Casey, J. L., 2010. Correlation of solar activity minimums and large magnitude geophysical events. *Space and Science Research Center, Research Report 1-2010* (Preliminary), p. 1–5.

Casey, J. L., 2012. Cold Sun. Trafford Publishing, 167p.

Casey, J. L., 2014. Dark winter: How the Sun is causing a 30-year cold spell. Humanix Books. 164p. www.Amazon.com.

Choi, D. R., 2010. The January 2010 Haiti seismic disaster viewed from the perspective of the energy transmigration concept and block tectonics. *NCGT Newsletter*(www.ncgt.org), v. 54, p. 36–44.

Choi, D. R., 2013a. An Archean geanticline stretching from the South Pacific to Siberia. *NCGT Journal*

(www.ncgt.org), v. 1, no. 3, p. 45–55.

Choi, D. R., 2013b. Earthquake/volcanic activities and solar cycles. *The Global Climate Status Report.* Edition 3-2013, September, p. 10-19. Space and Science Research Corporation. Orlando.

Choi, D. R., 2014. Seismo-volcanic energy propagation trends in the Central America and their relationship to solar cycles. *NCGT Journal*, v. 2, no. 1, p. 19–28.

Choi, D. R. and Maslov, L., 2010. Earthquakes and solar activity cycles. *NCGT Newsletter,* no. 54, p. 36–44.

Choi, D. R. and Tsunoda, F., 2011. Volcanic and seismic activities during the solar hibernation periods. *NCGT Newsletter,* no.61, p. 78–87.

Choi, D. R., Tsunoda, F. and Maslov, L., 2014. Seismo-volcanic energy propagation trends in the Aleutian Islands and North America. *NCGT Journal*, v. 2, no. 2, p. 13–22.

Greogry, G. P., 2002. Galaxy-Sun-earth relations. BeiträgezurGeoschichte der Geophysik und KosmischenPhysik, Band 3, Heft 4, 471p.

Iyengar, R. N., Sharma, D. and Siddiqui, J.M., 1999. Earthquake history of India in Medieval times. *Indian Journal of History of Science*, v. 34, no. 3, p. 181–237.

Jatskevich, B. A. (ed.), 2000. Geological Map of the World. 1:15,000,000 scale. Ministry of Natural Resources of Russian Federation, RAS.

Johnston, A.C. and Schweig, E.S., 1996. The enigma of the New Madrid Earthquakes of 1811-1812. *Anna. Rev. Planet. Sci.*, v. 24, p. 339–384.

Korhonen, J. V., Fairhead, J.D., Hamoudi, M, Hemant, K., Lesur, V., Mandea, M., Maus, S., Purucker, M. Ravat, D.,Sazonova, T. and Thebault, E., 2007. Magnetic anomaly map of the World (and associated DVD), Scale, 1:50,000,000, 1st edition, Commission for the Geological Map of the World, Paris, France.

Kosygin, Yu.A.,Basharin, A.K., Berzin, N.A., Borukayev, Ch.B., Matveyevskaya, A.L., Parfyonov, L.M., Chikov, B.M. and Schmidt, E.K., 1970. Structural and Material Complexes of the World. 1:15,000,000 scale. Compiled by Laboratory of Geotectonics, Institute of Geology and Geophysics, Siberian Branch, Academy of Scienceof USSR.

Meyerhoff, A.A., Taner, I., Morris, A.E.L., Agocs, W.B., Kamen-Kaye, M., Bhat, M.I., Smoot, N.C., Choi, D.R. and Meyerhoff-Hull, D. (ed.), 1996. *Surge tectonics: a new hypothesis of globalgeodynamics*. Kluwer Academic Publishers, 323p.

Nuttli OW. 1987. The effects of earthquakes in the central United States. *Rep. for Central US Earthq. Consort.*

Memphis TN: Fed. Emerg. Manage. Agency. 33p.

Pratsch, J.C., 2008. Letter to the Editor. *NCGT Newsletter*, no. 47, p. 4.

Pratsch, J.C., 2010. Gulf of Mexico Basin – a collapsed Late Carboniferous mantle dome? *NCGT Newsletter*, no. 55, p. 74–76.

Snyder, M., 2015. The New Madrid Earthquake that will divide the United States in half. 22 Feb. 2015.

http://endoftheamericandream.com/archives/the-new-madrid-earthquake-that-will-divide-the-united-states-in-half

Appendix 2, Item 6

USGS History of the 1811–1812 NMSZ Earthquakes[5]

New Madrid 1811–1812 Earthquakes

Earthquake Summary

Three Main Shocks

1. December 16, 1811 – Magnitude ~7.5
2. January 23, 1812 – Magnitude ~ 7.3
3. February 7, 1812 – Magnitude ~ 7.5

These had robust aftershock sequences. During the December 1811 earthquake, there were six aftershocks in the first two days in the range of M5.5 to M6.3. Hundreds of quakes were felt into 1813. *Aftershocks are earthquakes!*

December 16, 1811 - Magnitude ~7.0

It's happened before 1811–1812

[5] . USSG: http://earthquake.usgs.gov/earthquakes/states/events/1811-1812.php

The geologic record of pre-1811 earthquakes reveals that the New Madrid seismic zone has repeatedly produced sequences of major earthquakes, including several of magnitude 7 to 8, over the past 4,500 years. The New Madrid seismic zone is at significant risk for damaging earthquakes!

This map shows earthquakes (circles) of the New Madrid and Wabash Valley seismic zones (orange patches). Red circles indicate earthquakes that occurred from 1974 to 2002 with magnitudes larger than 2.5 located using modern instruments (University of Memphis). Green circles denote earthquakes that occurred prior to 1974 (USGS Professional Paper 1527). Larger earthquakes are represented by larger circles.

A Sequence of Three Main Shocks in 1811-1812

This sequence of three very large earthquakes is usually referred to as the New Madrid earthquakes, after the Missouri town that was the largest settlement on the Mississippi River between St. Louis, Missouri, and Natchez, Mississippi. On the basis of the large area of damage (600,000 square kilometers), the widespread area of perceptibility (5,000,000 square kilometers) and the complex physiographic changes that occurred, the New Madrid earthquakes of 1811–1812 rank as some of the largest in the United States since its settlement by Europeans. They were by far the largest east of the Rocky Mountains in the United States and Canada. The area of strong shaking associated with these shocks is two to three times as large as that of the 1964 Alaska earthquake and 10 times as large as that of the 1906 San Francisco earthquake. Because there were no seismographs in North America at that time and very few people in the New Madrid region, the estimated magnitudes of this series of earthquakes vary considerably and depend on modern researchers' interpretations of journals, newspaper reports, and other accounts of the ground shaking and damage. The magnitudes of the three principal earthquakes of 1811–1812 described below are the preferred values taken from research involved with producing the 2014 USGS National Seismic Hazard Map.

A Robust Aftershock Sequence for each Main Shock

The first principal earthquake, M7.5, occurred at about 2:15 a.m. (local time) in northeast Arkansas on December 16, 1811. The second principal shock, M7.3, occurred in Missouri on January 23, 1812, and the third, M7.5, on February 7, 1812 along the Reelfoot fault in Missouri and Tennessee. The earthquake ground shaking was not limited to these principal main shocks as there is evidence for a fairly robust aftershock sequence. The first and largest aftershock occurred on December 16, 1811 at about 7:15 a.m. At least three other large aftershocks are inferred from historical accounts on December 16 and 17. These three events are believed to range between M6.0 and 6.5 in size and to be located in Arkansas and Missouri. This would make a total of seven earthquakes of magnitude M6.0–7.5 occurring in the period from December 16, 1811 through February 7, 1812. In total, Otto Nuttli reported more than 200 moderate-to -large aftershocks in the New Madrid region between December 16, 1811 and March 15, 1812: ten of these were greater than about 6.0, about one hundred were between M5.0 and 5.9, and 89 were in the magnitude 4 range. Nuttli also noted that about 1,800 earthquakes of about M3.0 to 4.0 during the same period.

Large Area of Damaging Shaking

The first earthquake of December 16, 1811, caused only slight damage to man-made structures mainly because of the sparse population in the epicentral area. The extent of the area that experienced damaging earth motion, which produced Modified Mercalli Intensity greater than or equal to VII, is estimated to be 600,000 square kilometers. However, shaking strong enough to alarm the general population (intensity greater than or equal to V) occurred over an area of 2.5 million square kilometers.

Shaking Caused Sand Blows, River Bank Failures, Landslides, and Sunken Land

The earthquakes caused the ground to rise and fall—bending the trees until their branches intertwined and opening deep cracks in the ground. Deep-seated landslides occurred along the steeper bluffs and hillslides, large areas of land were uplifted permanently, and still larger areas sank and were covered with water that erupted through fissures or craterlets. Huge waves on the Mississippi River overwhelmed many boats and washed others high onto the shore. High banks caved and collapsed into the river, sand bars and points of islands gave way, and whole islands disappeared. Surface fault rupturing from these earthquakes has not been detected and was not reported, however. The region most seriously affected was characterized by raised or sunken lands, fissures, sinks, sand blows, and large landslides that covered an area of 78,000–129,000 square kilometers, extending from Cairo, Illinois, to Memphis, Tennessee, and from Crowley's Ridge in northeastern Arkansas to Chickasaw Bluffs, Tennessee. Only one life was lost in falling buildings at New Madrid, but chimneys were toppled and log cabins were thrown down as far distant as Cincinnati, Ohio, St. Louis, Missouri, and in many places in Kentucky, Missouri, and Tennessee.

A notable area of subsidence that formed during the February 7, 1812 earthquake is Reelfoot

Lake in Tennessee, just east of Tiptonville dome on the down-dropped side of the Reelfoot scarp. Subsidence there ranged from 1.5 to 6 meters although larger amounts were reported.

Other areas subsided by as much as 5 meters although 1.5 to 2.5 meters was more common. Lake St. Francis in eastern Arkansas, which was formed by subsidence during both prehistoric and the 1811–1812 earthquakes, is 64 kilometers long by 1 kilometer wide. Coal and sand were ejected from fissures in the swamp land adjacent to the St. Francis River, and the water level is reported to have risen there by 8 to 9 meters.

Large waves (seiches) were generated on the Mississippi River by seismically induced ground motions deforming the riverbed. Local uplifts of the ground and water waves moving upstream gave the illusion that the river was flowing upstream. Ponds of water also were agitated noticeably.

Surface Deformation: Evidence for Prehistoric Earthquakes

The Lake County uplift, about 50 kilometers long and 23 kilometers wide, stands above the surrounding Mississippi River Valley by as much as 10 meters in parts of southwest Kentucky, southeast Missouri, and northwest Tennessee. The uplift apparently resulted from vertical movement along several ancient subsurface faults. Most of the uplift occurred during prehistoric earthquakes. A strong correlation exists between modern seismicity and the uplift, indicating that stresses that produced the uplift may still exist today. Within the Lake County uplift, Tiptonville dome, which is about 14 kilometers in width and 11 kilometers in length, shows the largest up-warping and the highest topographic relief. It is bounded on the east by the 3-meter high Reelfoot scarp. Although most of Tiptonville dome formed between 200 and 2,000 years ago, additional uplifting deformed the northwest and southeast parts of the dome during the earthquakes of 1811–1812.

1811, December 16, 08:15 UTC Northeast Arkansas: The First Main Shock
2:15 a.m. local time
Magnitude ~7.5

This powerful earthquake was felt widely over the entire eastern United States. People were awakened by the shaking in New York City, Washington, D.C., and Charleston, South Carolina. Perceptible ground shaking was in the range of one to three minutes depending upon the observer's location. The ground motions were described as most alarming and frightening in places like Nashville, Tennessee, and Louisville, Kentucky. Reports also describe houses and other structures being severely shaken with many chimneys knocked down. In the epicentral area, the ground surface was described as in great convulsion with sand and water ejected tens of feet into the air (liquefaction).

1811, December 16, 13:15 UTC Northeast Arkansas: The "Dawn" Aftershock
7:15 a.m. local time
Magnitude ~7.0

A large event felt on the East Coast that is sometimes regarded as the fourth principal earthquake of the 1811–1812 sequence. The event is described as severe at New Bourbon, Missouri and was described by boatman John Bradbury, who was moored to a small island south of New Madrid, as "terrible, but not equal to the first." Hough believes that this large aftershock occurred around dawn in the New Madrid region near the surface projection of the Reelfoot fault.

1812, January 23, 15:15 UTC, New Madrid, Missouri
9:15 a.m. local time,
Magnitude ~7.3

This was the second principal shock of the 1811–1812 sequence. It is difficult to assign intensities to the principal shocks that occurred after 1811 because many of the published accounts describe the cumulative effects of all the earthquakes and because the Ohio River was iced over, so there was little river traffic and fewer human observers. Using the December 16 earthquake as a standard, however, there is a general consensus that this earthquake was the smallest of the three principals. The meizoseismal area was characterized by general ground warping, ejections, fissuring, severe landslides, and caving of stream banks.

1812, February 7, 09:45 UTC, New Madrid, Missouri
3:45 a.m. local time,
Magnitude ~7.5

This was the third principal earthquake of the 1811–1812 series. Several destructive shocks occurred on February 7, the last of which equaled or surpassed the magnitude of any previous event. The town of New Madrid was destroyed. At St. Louis, many houses were damaged severely and their chimneys were thrown down. The meizoseismal area was characterized by general ground warping, ejections, fissuring, severe landslides, and caving of stream banks.

*Abridged from *Seismicity of the United States, 1568–1989* (Revised), by Carl W. Stover and Jerry L. Coffman, U.S. Geological Survey professional paper 1527, United States Government Printing Office, Washington: 1993; "The enigma of the New Madrid earthquakes of 1811–1812" by A. C. Johnston, and E. S. Schweig in the *Annual Review of Earth and Planetary Sciences*, v. 24, p. 339–384 (doi: 10.1146/annurev.earth.24.1.339); "Cataloging the 1811–1812 New Madrid, Central U.S., Earthquake Sequence," by S. E. Hough, in the *Seismological Research Letters*, v. 80, no. 6, p 1045–1053 (doi: 10.1785/gssrl.80.6.1045).

Magnitudes updated based on 2014 seismic hazard mapping data.

Appendix 2, Item 7

List of Researchers Who Predict a New Cold Climate or Solar Hibernation

The following list represents only a tip-of-the-iceberg partial list of the many professionals around the world who believe a new cold climate or new ice age has begun or will soon begin.

1. **Dr. Habibullo I. Abdussamatov** of the Russian Academy of Scientists and head of space research at the Pulkova Observatory, St. Petersburg

Habibullo Abdusamatov said he and his colleagues had concluded that a period of global cooling similar to one seen in the late seventeenth century—when canals froze in the Netherlands and people had to leave their dwellings in Greenland—could start in 2012–2015 and reach its peak in 2055–2060. He said he believed the future climate change would have very serious consequences and that authorities should start preparing for them today.

Later, he said in his 2012 paper:

"We can expect the onset of a deep bicentennial minimum of total solar irradiance (TSI) in approximately 2042±11 and the 19th deep minimum of global temperature in the past 7500 years—in 2055±11. After the maximum of solar cycle 24, from approximately 2014 we can expect the start of deep cooling with a Little Ice Age in 2055±11."

2. **Jarl R. Ahlbeck, D.Sc.,** Åbo Akademi University, Finland

"Therefore, prolonged low solar activity periods in the future may cause the domination of a strongly negative AO and extremely cold winters in North America, Europe and Russia."

3. **SyunAkasofu, professor of geophysics**, emeritus, University of Alaska, also founding director of ARC

He predicts the current pattern of temperature increase of 0.5C /100 years resulting from natural causes will continue with alternating cooling as well as warming phases. He shows cooling for the next cycle until about 2030/ 2040.

4. **David Archibald, Summa Development Limited**, Australia

"Based on a solar maxima of approximately 50 for solar cycles 24 and 25, a global temperature decline of 1.5C is predicted to 2020 equating to the experience of the Dalton Minimum." (From his paper "Solar Cycles 24 and 25 and predicted climate response," *Energy and Environment*, vol.17, no.1., Archibald, D.C.:, 2006)

From his 2014 book, *Twilight of Abundance*, he says "The Sun drives climate. The demonstrated relationship between solar activity and climate predicts a severe cooling out to at least the year 2040—that is, for the next quarter of a century or so."

5. **Dr. O.G.Badalyan and Dr.V.N. Obridko, Institute of Terrestrial Magnestism. Russia; Dr.J.Sykora, Astronomical Institute of the Slovak Academy of Sciences, Slovak Republic**

"A slow increase in (intensity of coronal green line) in the current cycle 23 permits us to forecast a low-Wolf-number (number of sunspots) cycle 24 with the maximum W~50 at 2010-2011." (From their paper "Brightness of the coronal green line and prediction for activity cycles 23 and 24," *Solar Physics,* 199: pp.421–435.)

Note that a 50 sunspot level is a Dalton-class minimum.

6. **Dr. Karsten Brandt,** director of donnerwetter.de weather service

"It is even very probable that we will not only experience a very cold winter, but also in the coming 10 years every second winter will be too cold. Only 2 of 10 will be mild."

7. **Dr. B. P. Bonev, Dr. Kaloyan M. Penev, Dr. Stefano Sello**

"We conclude that the present epoch is at the onset of an upcoming local minimum in long term solar variability."(From their paper "Long term solar variability and the solar cycle in the 21st century," *The Astrophysical Journal, vol. 605*, pp.L81–L84.)

8. **John L. Casey, director, Space and Science Research Center, Orlando, Florida**

From the center's research report, "The existence of 'relational cycles' of solar activity on a multi-decadal to centennial scale, as significant models of climate change on earth – The RC Theory," (www.spaceandscience.net) John L. Casey says, "As a result of the theory, it can be predicted that the next solar minimum may start within the next 3–14 years, and last 2–3 solar cycles or approximately 22–33 years . . . It is estimated that there will be a global temperature drop on average between 1.0 and 1.5 degrees C, if not lower, at least on the scale of the Dalton Minimum . . . This forecast next solar minimum will likely be accompanied by the coldest period globally for the past 200 years and as such, has the potential to result in worldwide, agricultural, social, and economic disruption."

9. **Peter Clark, professor of Geosciences** at OSU

"Sometime around now, scientists say, the Earth should be changing from a long interglacial period that has lasted the past 10,000 years and shifting back towards conditions that will ultimately lead to another ice age."

10. **Piers Corbyn, astrophysicist**

Corbyn suggested we should sooner prepare for another ice age than worry about global warming. Corbyn believed global warming "is complete nonsense, it's fiction. It comes from a cult ideology. There's no science in there, no facts to back [it] up."

11. **Joe d'Aleo, executive director of Certified Consultant Meteorologists**

"Longer term the sun is behaving like it did in the last 1700s and early 1800s, leading many to believe we are likely to experience conditions more like the early 1800s (called the Dalton Minimum) in the next few decades. That was a time of cold and snow. It was the time of Charles Dickens and his novels with snow and cold in London."

12. **Dr. Don Easterbrook, professor emeritus, Department of Geology, Western Washington University.**

"Setting up of the PDO cold phase assures global cooling for next approx. 30 years. Global warming is over. Expect 30 years of global cooling, perhaps severe 2–5°F."

13. **Dr. Alexander Frolov, head of Russia's state meteorological service Rosgidromet**

"From the scientific point of view, in terms of large scale climate cycles, we are in a period of cooling . . . The last three years of low temperatures in Siberia, the Arctic and number of Russia mountainous regions prove that, as does the recovery of ice in the Arctic Ocean and the absence of warming signs in Siberia."

14. **Dr. William M. Gray, professor emeritus, Department of Atmospheric Sciences, Colorado State University**

"A weak global cooling began from the mid-1940's and lasted until mid-1970's. I predict this is what we will see in the next few decades."

15. **Thomas Globig, meteorologist,** says in 2010:

"It is quite possible that we are at the beginning of a Little Ice Age."

16. **Dr. Fred Goldberg, Swedish climate expert** says

"We could have an ice age at any time."

17. **Dr. Peter Harris,** engineer, retired, Queensland, Australia

"We can say there is a probability of 94% of imminent global cooling and the beginning of the coming ice age."

18. **Victor Manuel Velasco Herrera**, researcher at the National Autonomous University of Mexico

The following are his comments from his research released in August 2008: "In two years or so, there will be a small ice age that lasts for 60–80 years."

19. **Drs. Y. T.Hong, H. B. Jiang, T. S. Liu, L .P.Zhou, J. Beer, H. D. Li, X. T. Leng, B .Hong,** and **X.G. Qin**

From their paper "Response of climate to solar forcing recorded in 6,000-year (isotope) O18 time-series of Chinese peat cellulose," in *The Holocene 10.1* (2000), pp. 1–7:

The Chinese team of researchers observed "a striking correspondence of climate events to nearly all of the apparent solar activity changes."

In showing O18 isotope measurements were high during the coldest periods, they concluded, "If the trend after AD 1950 continues . . . the next maximum of the peat O18 (and therefore cold maximum) would be expected between about AD 2000 and AD 2050."

20. **Dr. Boris Komitov, Bulgarian Academy of Sciences, Institute of Astronomy,** and **Dr. Vladimir Kaftan, Central Research Institute of Geodesy, Moscow**

From their paper "The sunspot activity in the last two millennia on the basis of indirect and instrumented indexes: time series models and their extrapolations for the 21st century," presented at the International Astronomical Union Symposium No. 223, they say, "It follows from their extrapolations for the 21st century that a supercenturial solar minimum will be occurring during the next few decades . . . It will be similar in magnitude to the Dalton minimum, but probably longer as the last one."

21. **Dr. George Kukla, Czechoslovakian Academy of Sciences** says

"In the 1970's leading scientists claimed that the world was threatened by and era of global cooling. . . Based on what we have learned this decade, those scientists . . . had it right. The world is about to enter another Ice Age."

22. **Dr. Theodor Landscheidt** (1927–2004), Schroeter Instiitute for Research in Cycles of Solar Activity, Canada

Among his comments from many years of research on solar climate forcing include "Contrary to the IPCC's speculation about man made warming as high as 5.8(degrees)C within the next hundred years, a long period of cool climate with its coldest phase around 2030 is to be expected."

23. **Prof. Mojib Latif, Professor**, Kiel University, Germany says

"You may well enter a decade or two of cooling relative to the present temperature level."

24. **Dr. Grima Orssengo says**

"These cool and warm PDO regimes correlate well with the cooling and warming phases…the model…predicts cooling until 2030."

25. **Dr. Tim Patterson, Department of Earth Sciences, Carleton University, Canada**

From an article in the *Calgary Times*, May 18, 2007. Indeed, one of the more interesting, if not alarming, statements Patterson made before the Friends of Science luncheon is that satellite data shows that by the year 2020, the next solar cycle is going to be solar cycle 25—the weakest one since the Little Ice Age (that started in the thirteenth century and ended around 1860), a time when people living in London, England, used to walk on a frozen Thames River and food was scarcer. Patterson said, "This should be a great strategic concern in Canada because nobody is farming north of us." In other words, Canada—the great breadbasket of the world—just might not be able to grow grains in much of the prairies.

26. **Dr. Oleg Pokrovsky**, Voeikov Main Geophysical Observatory:

Ria Novosti writes, "There isn't going to be an ice age, but temperatures will drop to levels last seen in the 1950s and 1960s . . . Right now all components of the climate system are entering a negative phase . . . The cooling will reach its peak in 15 years. Politicians who have geared up for warming are sitting on the wrong horse . . . The Northeast Passage will freeze over and will be passable only with icebreakers."

27. **Dr. Nicola Scafetta, Duke University**

"Empirical evidence for a celestial origin of the climate oscillations and its implications . . . The partial forecast indicates that climate may stabilize or cool until 2030-2040."

28. **Drs. Ken K. Schatten and W. K.Tobiska**

From their paper presented at the Thirty-Fourth Solar Physics Division meeting of the American

Astronomical Society, June 2003: "The surprising result of these long range predictions is a rapid decline in solar activity, starting with cycle #24. If this trend continues, we may see the Sun heading towards a 'Maunder' type of solar activity minimum—an extensive period of reduced levels of solar activity."

29. **Dr. Oleg Sorokhtin**, merited scientist of Russia and fellow of the Russian Academy of Natural Sciences and researcher at the Oceanology Institute

From recent news articles regarding the next climate change, he has said, "Astrophysics know two solar cycles, of 11 and 200 years. Both are caused by changes in the radius and area of irradiating solar surface . . . Earth has passed the peak of its warmer period and a fairly cold spell will set in quite soon, by 2012. Real cold will come when solar activity reaches its minimum, by 2041,and will last for 50–60 years or even longer."

30. **Professor Anastasios Tsonis**, head of Atmospheric Sciences Group University of Wisconsin, and **Dr. Kyle Swanson** of the University of Wisconsin-Milwaukee

"We have such a change now and can therefore expect 20–30 years of cooler temperatures . . . This is nothing like anything we've seen since 1950."

31. **Drs. Ian Wilson, Bob Carter, and I. A. Waite**

From their paper "Does a Spin-Orbit Coupling Between the Sun and the Jovian Planets Govern the Solar Cycle?" in the *Publications of the Astronomical Society of Australia*, 25(2) 85–93, June 2008, Dr. Wilson clarifies, "It supports the contention that the level of activity on the Sun will significantly diminish sometime in the next decade and remain low for about 20-30 years. On each occasion that the Sun has done this in the past the World's mean temperature has dropped by ~ 1–2° C."

32. **Drs. Lin Zhen-Shan and Sun Xian**, Nanjing Normal University, China

From their paper in *Meteorology and Atmospheric Physics*, 95,115–121 "Multi-scale analysis of global temperature changes and trend of a drop in temperature in the next 20 years": "We believe global climate changes will be in a trend of falling in the following 20 years."

33. **Dr. Valentina Zharkova, et al.,** University of Northumbria, England,

"[In the cycle between 2030 and around 2040] the two waves exactly mirror each other—peaking at the same time but in opposite hemispheres of the sun," she said. "Their interaction will be disruptive, or they will nearly cancel each other . . . We predict that this will lead to the properties of a 'Maunder minimum . . . We found that our predictions showed an accuracy of 97 per cent."

The list above was compiled from a list at the Space and Science Research Corporation (2014) and from Matt Voora, PE (2010).

Appendix 2, Item 8

Chronology of Notifications to Government, Media, and the Public of New Cold Climate and Associated Catastrophic Earthquakes and Volcanic Eruptions

1. March 2, 2010. The SSRC issues its second Research Report 1-2010 (Preliminary) and associated press release. The report is titled "Correlation of Solar Activity Minimums and Large Magnitude Geophysical Events."

   This research report provides substantial evidence of the likelihood of major, possibly historic volcanic eruptions and earthquakes occurring during the ongoing solar hibernation. In this report, a high probability is established for major geophysical events based upon analysis of the last 400 years of the largest earthquakes in the United States and volcanic eruptions worldwide. The associated press release is titled "Sun's Activity Linked to Largest Earthquakes and Volcanoes."

   The important findings from this report are discussed in chapter 2 "The Sun-Earth Connection."

Note: Almost one year later to the day, after issuing this press release, on March 11, 2011, a devastating M9.0 earthquake and subsequent tsunami killed over 15,600 in Japan and caused the Fukushima Daiichi nuclear plant to have three of four operating reactors to explode/meltdown.

See the detailed Research Report 1-2010 in appendix 4 and Press Release SSRC 1-2010 in appendix 2.

2. March 14, 2011. The SSRC issues Press Release 4-2011, warning that there will be more and large earthquakes like that which struck Japan on March 11, 2011.

   The SSRC reinforces its March 2010 prediction for historic earthquakes and volcanic eruptions during the current solar hibernation. See the Press Release 4-2011 in appendix 2.

3. In the fall of 2011, my first book on climate variation, which was endorsed by other scientists, *Cold Sun*, was published by Trafford Publishing (Penguin). The book lays out the case for the relationship between solar cycles and record earthquakes and volcanic eruptions. Cited in the book were the examples of the New Madrid series of earthquakes, the largest ever in the United States that took place from December 1811 to February 1812. These temblors ranged from M7.5 to M8.0 and caused the Mississippi River to flow backward briefly. Church bells rang in New England because of the power of this event. The book also indicates that the largest ever recorded volcanic eruption took place at Mt. Tambora in Indonesia in April 1815. Both these record events happened at the bottom of the last solar hibernation during the so-called Dalton minimum from 1793 to 1830.

4. February 2, 2012. The International Earthquake and Volcano Prediction Center (IEVPC) is formed. In a press release, the IEVPC discusses its plans to develop a process for earthquake prediction using multiple precursor signals.

5. October 2012. The IEVPC begins its first test program to determine the effectiveness of its earthquake-prediction system of integrated precursor signals. During the process, the IEVPC notifies various foreign governments of the likelihood of earthquakes based on the IEVPC process.

6. December 2012. The IEVPC completes its first test program and announces its highly successful results in a press release. See the PR at appendix 2.

7. June 2013, the IEVPC begins a second one-year-long test program while temporarily suspending notification of predictions for respective countries.

8. September 13, 2013. The SSRC publishes the commentary/research paper titled "Earthquake/ volcanic activities and solar cycles" by Dr. Choi in its Global Climate Status Report (GCSR) Edition 3-2013. In this paper, Dr. Choi lays out in substantial detail a strong correlation between the strongest, most damaging earthquakes and volcanoes and solar activity. See the complete paper in appendix 4. The original paper was published as "Earthquakes and Solar Activity Cycles" in the *NCGT Newsletter*, no. 57, p. 85–97. Much of that paper was reproduced in the GCSR commentary by Dr. Choi and in a geological journal, *New Concepts in Global Tectonics* (NCGT), published by Dr. Choi and reviewed by a team of international geologists. I believe that the 2010 paper by Dr. Choi and Dr. Maslov was the first time that an unmistakeable connection between the Sun and earthquakes was established. This inverse relationship is what I call the Choi relationship. It was subsequently validated by a joint paper by Dr. Choi and Dr. Tsunoda, professor emeritus, National Saitama University, Japan.

9. March 2014. A summary research paper is authored by Dr. Choi, Dr. Maslov, Dr. Tsunoda, and Casey, which was published in the SSRC GCSR Edition 1-2014. This summary cites previous research of the authors with the following conclusion:

   "The increasing seismic activity since 1990 is expected to continue for the coming two to three decades as we have entered a 'solar hibernation' or possibly a mini-ice age (Casey, 2012); this will likely bring more strong, possibly catastrophic earthquakes."

10. May 29, 2014. IEVPC initiates insurance industry contacts to try to get them and their individual and business clients up to speed on the risk of coming earthquakes. The IEVPC contacts major U.S. national insurance companies. Another reason for the initiative includes "the use of U.S. and foreign government channels has shown that it will take too long to change long standing beliefs about earthquake prediction."

11. April 28, 2014. The president is notified for the *final* time to prepare the nation for the coming solar hibernation via a letter and widely publicized Press Release 2-2014. He and his staff had been routinely made aware of the coming cold climate by me via email and letters since his presidential campaign in 2008. See the PR 2-2014 and letter at appendix 2.

In the letter among other items was the following:

"Research related to these solar hibernations, also shows they occur concurrently with the most destructive earthquakes and volcanic eruptions, the latter of which can add dramatically to an already colder climate."

A typical, though depressing, written response was actually received later as a form letter with a computerized signature of the president, extolling the virtues of the president's attempts to control man-made global warming.

12. July 2014. The IEVPC completes its second test program. Based upon the success of the first two test programs, the IEVPC resumes earthquake predictions and posting of earthquake warnings at its web site.

13. June 2014. Humanix Books publishes an updated, restructured version of *Cold Sun* as *Dark Winter* and covered the correlation between record earthquakes and volcanic eruptions with solar hibernations.

14. 2014–2015. Hundreds of radio and NewsMax TV interviews and public presentations were conducted by John Casey, alerting millions of Americans of the need to prepare for the coming cold climate with its associated major earthquakes and volcanic eruptions.

15. Late July, early August 2015. The book *Dark Winter* becomes the number 1 bestselling climate book at Amazon.com. It also becomes the only book sold on Amazon.com during 2015 to have also been number 1 bestselling book in the other categories of public policy, astronomy and astrophysics, Earth sciences, and weather.

16. August 25, 2014. The IEVPC posts and distributes its press release titled "California Enters Greatest Earthquake Risk Period." The release is sent to the Office of the Governor of California and major newspapers in the state. The detailed press release explains the solar activity connection with major earthquakes and the 206-year cycle of the Sun that is also causing a shift into a new cold climate.

17. June 5, 2015. The SSRC sends a letter to FEMA Administrator Craig Fugate with a warning that the United States will enter its highest risk period for catastrophic earthquakes and volcanic eruptions during the period of 2017 to 2038. In the letter, the SSRC emphasizes

the correlation between climate change, solar hibernations, and record quakes and volcanoes. No response to this letter was ever received even though it was from a climate research organization with the best public track record for climate prediction in the United States. A call was then placed to the FEMA administrator's office after a month. The FEMA administrator's office said they did not believe the June letter required any action!

18. June 2015. Numerous mainstream media outlets and prominent state newspapers were notified via email of the June 5, 2015 FEMA letter. No reply was received from anyone.

19. June 2015. During the month of June, the SSRC notified each applicable governor's office of the increased threat of record quakes and volcanic eruptions vis-à-vis the letter to FEMA Administrator Fugate. All West Coast states and central Mississippi states and South Carolina were notified. No meaningful reply that resulted in follow-on contact was received.

20. September 2015. NewsMax Media Inc., parent of Humanix Books, begins to air a TV documentary about the book *Dark Winter* that discusses the coming cold climate and associated destructive earthquakes and volcanoes. Millions view it on Dish Network, Direct TV, Verizon FiOS, and online at NewsmaxTV.com. The documentary was still showing as of late June 2016, segments of which have since been seen copied on to the internet.

21. May 13, 2016. A follow-up letter and additional warning was sent to the FEMA administrator, indicating that the June 5, 2015 letter from the SSRC should be taken more seriously in view of recent geophysical activity and the warning issued by Dr. Thomas Jordan at the Southern California Earthquake Center on May 4, 2016. On May 24, 2016, I called FEMA to see whether a reply was forthcoming. A perturbed person on the other end of the phone with a "how dare you" tone in her voice said the letter was still being "scanned" and had not been given to the FEMA administrator yet. They indicated the IEVPC would be notified if any actions were to be taken. On July 5, 2016, I received a response to the May 13 letter to the FEMA administrator. It was the typical government no-response response. It said things like "FEMA recognizes the catastrophic nature of these disasters and is working with the state, local, tribal, and territorial governments to ensure citizens are better prepared to respond to disasters." My guess is that the FEMA reply was a standard form letter given out to many others who have expressed concerns over predicted earthquakes. FEMA just had to change the addressee to accommodate my concerns and wash its hands of the matter. Like the June 5, 2015 letter, no one at FEMA ever contacted the SSRC or the IEVPC

or me directly about these warnings to inquire of the research behind them or even ask a simple question of any kind!

22. May 25, 2016. A second mailing is sent out to the governors of the U.S. states most likely to suffer the brunt of coming geophysical disasters. It included the two previous letters sent to the FEMA administrator. These states include Alaska, Washington, Oregon, California, Missouri, Illinois, Indiana, Kentucky, Tennessee, Mississippi, Arkansas, and South Carolina. On July 5, 2016, additional letters were sent to the governors of Alabama, Hawaii, and Puerto Rico. See a sample letter to one of the governors in appendix 2. At no time has any state governor's office contacted me to discuss the IEVPC's research that has led to the conclusions that their state should prepare for catastrophic geophysical events.

23. June 6, 2016. The IEVPC posts and distributes its first press release of the year, "Federal and State Leaders Warned to Prepare for Catastrophic Earthquakes and Volcanoes." This release, as indicated in the text, represents the "final warning" to governments after years of prior warnings to get their states and federal agencies into immediate preparation mode. The high-risk areas of the U.S. West Coast, the central Mississippi valley, and South Carolina were singled out for immediate preparation with the expectation that catastrophic quakes could strike anytime but definitely within the period of 2017 to 2038. The governor of Alaska and Puerto Rico were also later notified. Most of the top twenty U.S. newspapers were contacted as well as the U.S. Senate and U.S. House of Representative science committees over the following two days.

    During spring of 2016, in total, hundreds of emails and letters were sent out to the media, state and federal government officials, and other important individuals warning of the need to make immediate preparations in view of the high probability of catastrophic earthquakes and volcanic eruptions. The high-risk window of 2017 to 2038 was spelled out. There was not one positive response from any of these many notifications asking for further information or a statement that they were going to act on the recommendations to warn the public and otherwise prepare!

****

As matter of course, over the years, both the SSRC and IEVPC widely distributed their press releases to include most major media outlets, TV and print, along with leadership in Congress as well as of Senate and House science committees. Prior notifications of the coming cold climate and its potentially dangerous ill effects were sent to every U.S. senator twice, each state governor (twice for many of them), and every attorney general. A host of special communications has been sent to the White House, key agencies like NOAA and NASA, DOD, FBI, USDA, and other department heads during the years 2007 to 2016 from either the SSRC or the IEVPC.

Since the current administration began in 2008, there has been no serious reply or request for further information from a federal office, including requests for the research papers used to back up the recommendations for preparing for the new cold climate or concurrent major earthquakes and volcanic eruptions.

This regrettable result comes with the fact that during its time, the SSRC had one of the most successful track records of climate prediction of any research organization in the USA that the book covering this subject, *Dark Winter*, was a 2015 number 1 bestseller on Amazon.com, with a TV documentary seen by millions. In addition, the IEVPC, made up of a distinguished team of international scientists, had one of if not the best public records of earthquakes predictions of any private science organization in the USA and perhaps globally.

Appendix 2, Item 9

Press release: International Earthquake and Volcano Prediction Center

P.O. Box 607147 - Orlando, Fl 32860
(407) 985-3509 - mail@ievpc.org

Earthquake Prediction Center Ends Successful Test Program Early

Tuesday, December 18, 2012 Press Release 03-12-18-12
8:00 AM EST

The recently organized International Earthquake and Volcano Prediction Center (IEVPC) announces today that it has stopped its internal earthquake test program early because of a near perfect record in its predictions and the vital need to begin saving lives immediately.

The IEVPC has just achieved an almost flawless level in earthquake prediction as demonstrated in the first three tests of its Catastrophic Geophysical Event (CGE) Monitoring and Warning System (CMWS). As a result, the IEVPC has decided to stop further evaluations and immediately begin notification of governments around the world of its now verified ability to predict large destructive earthquakes with a high degree of certainty.

According to Chairman/CEO John Casey, "We can no longer hold back in letting the earthquake prone nations of the world know that a proven system for highly reliable prediction of large earthquakes now exists. These geophysical dangers routinely kill thousands of people around the world every year, while at the same time destroying homes, businesses, and infrastructure, thereby extending the damage and suffering for many thousands more. The need to cut short our internal test program, originally planned for almost twenty earthquakes, is obvious. We now have a process for earthquake prediction that is so reliable that it must immediately be put into place wherever lives are at risk. The decision we have made to stop evaluation of our CMWS is similar to important drug testing programs. It is not unusual for promising new drugs to have testing stopped abruptly if the initial results are so compelling and people are dying every day without the drug. Likewise, we have decided we must not wait any longer but must aggressively get out the word about our capabilities.

What we need now is for nations of the world to recognize that it is a myth that earthquakes cannot be predicted and to begin to establish communication networks and standardized monitoring systems in known high risk zones. Other international groups are also coming out with effective tools for earthquake prediction. They, like us, realize that CGE's can be predicted because of recent advances in technology, especially satellite sensor technology, and because of the integration of many prediction techniques and precursor signals into a single predictive on techniques and precursor signals into a single predictive process. Once in place, we believe we can maximize the time people have to prepare for these destructive events by providing months, weeks, and days of advance notice. While we will doubtless continue to improve our process for quake detection, there is no longer a need to continue the test program. At the same time, there is an overwhelming humanitarian need to end it.

In September of last year I was approached by some of the world's best seismologists in earthquake prediction to create this new organization that would integrate their combined skills, techniques, and decades of experience. They came to me because of my success in climate change prediction and especially how it relates to variations in earthquake and volcanic activity. Under the leadership of Director of Research Dr. Choi, we have been busy assembling the best and brightest in earthquake prediction under one roof. That effort has now paid off. Many people worldwide will ultimately benefit from this initiative."

In the past two months, the IEVPC concluded three separate tests in different areas of the world included the following:

1. Kamchatka Peninsula, Russia. This test resulted in the correct prediction of timing and location of a major earthquake event that resulted in an amazing ten earthquakes

ranging from M4.6 to M5.8 over a short eight day period spread along a fault line to a distance of about 1,500 km. Eight of the quakes hit within the first two days. These temblors, in combination, replaced the IEVPC's previously predicted single quake of M7.5–M8.8. Mercifully for the people of Kamchatka and the Pacific Rim, as the substantial energy of the quake being monitored by the IEVPC approached the surface, it dispersed among several faults lines off the east coast of Kamchatka during October 14–22, 2012. This quake in its final form of multiple powerful quakes produced no known loss of life. Had a single quake struck, thousands of lives might have been lost because of direct quake effects and the generation of a Pacific-wide tsunami. Several IEVPC Associate Scientists were involved with this prediction including Dr. Z. Shou, Dr. M. Hayakawa, Dr. A. Bapat and V. Straser under the leadership of IEVPC Director of Research, Dr. Dong Choi.

2. Celebes Sea of Northern Indonesia. As a result of IEVPC precursor analysis conducted by lead investigator and Director of Research, Dr. Choi, a large oceanic quake (M6.0) was correctly predicted and took place on October 17, 2012 at the location and within the time frame estimated. Because of the deep ocean nature of this isolated quake's epicenter, no damage or loss of life was recorded.

3. Myanmar. On November 11, 2012 a M6.8 quake struck central Myanmar near the location predicted with the magnitude and in the time frame as internal IEVPC estimates had forecast. Twelve lives were lost based on initial figures released by the government.

   Leading the initial precursor signal analysis and early detection of the Myanmar quake was renowned Indian seismologist Dr. Arun Bapat. Dr. Choi was also involved in this quake's analysis and used other signals to confirm Dr. Bapat's preliminary conclusions. The final opinion arrived at was for a potentially catastrophic geophysical event (CGE) which would strike central Myanmar within two weeks after November 6, 2012 and would have a magnitude between M6.5 and M7.0.

In his assessment of the Myanmar test CEO Casey explained, "I am of course delighted at the success of Dr. Bapat and Dr. Choi in their trial prediction of the Myanmar quake. Dr. Bapat is one of the most distinguished leaders in this field and his history making prediction of the Myanmar quake is only one example of the talent that resides in the IEVPC. Unfortunately, it was impossible to know whether the Myanmar quake would happen since the IEVPC process had not been evaluated for an inland quake before. Issuing a warning was out of the question for what was then an unproven method with such a short timespan to strike, in a country with little or no effective earthquake reaction training for its citizens. It would have been grossly irresponsible. Issuing a public alert ran the risk of possibly causing panic throughout the country, resulting in far more deaths than that seen in the remote areas where the quake epicenter was located."

From Dr. Bapat we have, "The fact that the Myanmar quake struck as predicted, and that it did so along with the Celebes Sea quake and the Kamchatka quake event has given us enough justification to end the test program early. Mr. Casey and Dr. Choi have done a great service to all by asking those like myself with many years in the field of earthquake prediction to come together to end the myth that these destructive earthquakes cannot be predicted. I believe we are now at that point in human history."

Dr. Choi added, "We have had a remarkable level of success in our very first three tests. Further, they included diverse geophysical situations. The Kamchatka event was a traditional off shore Pacific Rim oceanic trench fault type. The Celebes Sea quake was a central oceanic deep ocean event with no companion fault. The Myanmar event was an inland quake with an associated known fault line.

What is important to note is that our process worked correctly in three distinctly different geological areas. This gave us another reason for ending the test program quickly. If our process had worked only for one type of quake and not others we might have had to stop and reevaluate our process. That is no longer required. It's time to put our program for earthquake prediction in the field and start saving lives."

Mr. Casey echoes Dr. Choi's comments with, "This level of success in our predictions for Kamchatka, the Celebes Sea and Myanmar carries even more significance when one realizes all our work has been done in a start-up phase on a shoestring budget. A greater level of prediction success and improved warning notification time can be achieved for a state, region, or nation with requisite funding of global and on-site monitoring teams from the IEVPC.

"In the special case of Kamchatka where we issued warnings during much of 2012, we remain concerned for the potential of another major seismic event and all should remain vigilant in that unique region of the planet where there is a history of powerful earthquakes. Our Russian colleagues are, however, well versed in parametric precursor analysis similar to what we employ. We have also provided them additional information to detect a new unexpected earthquake and quickly react should that highly unstable area produce another threat. In any case it was gratifying to see the positive level of reaction to our warnings demonstrated by Russian geologists, Ambassador Kislyak's office in Washington, and by President Medvedev's trip to Kamchatka in August. While there he checked on the status of earthquake preparedness. We will nonetheless continue to keep Kamchatka on an active but lower alert status over the next year until relative stability returns. We believe a major earthquake will remain a serious threat for Kamchatka residents for some time.

"Beginning this week we will start a systematic program for raising the capital needed to expand our operations out of our start-up phase and notify every nation that has to deal with CGE's that we are here and able to help protect their people. The IEVPC has demonstrated that a new era in reliable earthquake forecasting has arrived."

****

The International Earthquake and Volcano Prediction Center is headquartered in Orlando, Florida, USA. The primary research facility is in Canberra, Australia, with initial branch offices of cooperating scientists and researchers planned for the USA, India, China, and Japan. The IEVPC is a non-profit science research organization dedicated to the mission of protection of people through early prediction of catastrophic geophysical events (CGE) such as earthquakes, associated tsunamis, and volcanic eruptions.

The IEVPC web site is at www.ievpc.org.

Edit: As of September 18, 2016, the new IEVPC phone number is 407-601-3295.

Appendix 2, Item 10

Seismo-Volcanic Energy Propagation Trends in the Aleutian Islands and North America

(Published in the *New Concepts in Global Tectonics (NCGT) Journal*, vol.2, no.2, Jun. 2014)

Dong Choi,[6] Fumio Tsunoda,[7] and Leo Maslov[8]

Abstract: Analysis of the time-space distribution of major earthquakes and volcanic eruptions in the Pacific coast of the Aleutians and North America allows to deduce the seismo-volcanic energy flow trends; westward from Alaska to Aleutian, and southward from Southern Alaska, via California, to Mexico. The average rate of earthquake epicentre propagation along a trend line in the shallow mantle above 50 km is consistent in all regions; 120–140 m/day (45–50 km/year). Whereas the speed of volcanic eruption propagation trend was calculated to be 470–620 m/day (170–225 km/year), much faster than that of seismic energy, probably due to the deeper root of volcanoes (100 to 250 km). In addition, an eastward counter movement (67 km/year =185 m/day) is also detected in both volcanic and earthquake distribution in the Aleutian region, implying the presence of a complex system of flow channels at different depths. Seismo-tomographic images of the whole mantle suggest that the seismo-volcanic

6 . International Earthquake and Volcano Prediction Center, Australia. dchoi@ievpc.org.
7 . Saitama University (Emer. Prof.), Japan. Tsunochan2@gmail.com
8 . Department of Physics and Astronomy, University of Northern Colorado, USA. lev.maslov@unco.edu

energy in the study region originates from the superplume in the South Pacific, crosses the Pacific through Hawaii in the middle mantle, and emerges in the Gulf of Alaska in the shallow mantle. The thermal gradient is considered to drive the flow to the west and the south along the mobile tectonic belts and deep fracture zones developed in the coastal margins of the Aleutians and North America. The effect of Earth rotation and other factors on the energy flow direction and speed are yet to be clarified.

Keywords: *Aleutians, Alaska, North America, earthquake, volcano, propagation speed*

1. Introduction

Earlier, we (Tsunoda et al., 2013) reviewed the current understanding of the thermal energy transmigration and fluctuation. The study clarified thermal energy flow mainly in the western Pacific and in African superplume provinces. This paper is a follow up of our previous work with a focus on the energy propagation observed in the North American earthquakes and volcanic eruptions.

Latest studies by us (Choi, 2014; Tsunoda and Kawabe, 2014, for example) further confirmed the energy flow in the lithosphere through the Circum Pacific Ceno-Mesozoic mobile belts and deep fault zones.

Today the energy flow concept originally proposed by Blot (1976) and developed by Meyerhoff et al. (1994) and Tsunoda et al. (2013) is essential in understanding the mechanism of great earthquakes and volcanic eruptions, which can be widely applied to practical fields of science—one example being the successful prediction of the March 2014 M6.7 Tarapaca Earthquake, Northern Chile, which was follow by a gigantic M8.2 mainshock two weeks later (Choi, 2014).

2. Energy flow deduced from the time-space distribution of strong earthquakes and volcanic eruptions

1) Aleutian to Alaska

The Time-space distribution of very strong earthquakes (M6.5+) and volcanic eruptions (VEI 3+) were extracted from the USGS (http://earthquake.usgs.gov/earthquakes/search/) and

IRIS (www.iris.edu/seismon/), and Smithsonian Institution Global Volcanic Program (http://volcano.si.edu/reports_weekly.cfm) for this study. They are listed in **Tables 1 and 2.**

Table 1. Strong earthquakes (M6.5+), Alaska – Aleutian, from 1970 to 2014 (April). M7.0+ quakes are highlighted.

Magnitude	(km)	Year	Day	Time	Latitude	Longitude	Locality
6.6	11.4	2014	24/04/2014	3:10:12	49.85	-127.44	VANCOUVER ISLAND REGION
7	33.5	2013	30/08/2013	16:25:02	51.61	-175.36	ANDREANOF ISLANDS, ALEUTIAN IS.
7.5	10	2013	5/01/2013	8:58:19	55.39	-134.65	SOUTHEASTERN ALASKA
7.8	14	2012	28/10/2012	3:04:08	52.79	-132.1	QUEEN CHARLOTTE ISLANDS REGION
6.5	22	2011	9/09/2011	19:41:34	49.54	-126.89	VANCOUVER ISLAND REGION
6.8	32	2011	2/09/2011	10:55:53	52.17	-171.71	FOX ISLANDS, ALEUTIAN ISLANDS
7.3	52	2011	24/06/2011	3:09:39	52.05	-171.84	FOX ISLANDS, ALEUTIAN ISLANDS
6.5	45.8	2010	3/09/2010	11:16:08	51.63	-176	ANDREANOF ISLANDS, ALEUTIAN IS.
6.7	21.5	2010	18/07/2010	5:56:45	52.75	-169.77	FOX ISLANDS, ALEUTIAN ISLANDS
6.6	8.7	2009	17/11/2009	15:30:46	51.96	-131.6	QUEEN CHARLOTTE ISLANDS REGION
6.6	5.1	2008	2/05/2008	1:33:35	51.86	-177.49	ANDREANOF ISLANDS, ALEUTIAN IS.
6.6	26.4	2008	16/04/2008	5:54:21	51.88	-179.1	ANDREANOF ISLANDS, ALEUTIAN IS.
6.5	25.9	2008	15/04/2008	22:59:54	51.9	-179.37	ANDREANOF ISLANDS, ALEUTIAN IS.
6.6	10	2008	5/01/2008	11:01:05	51.26	-130.76	QUEEN CHARLOTTE ISLANDS REGION
7.2	36.2	2007	19/12/2007	9:30:28	51.4	-179.54	ANDREANOF ISLANDS, ALEUTIAN IS.
6.5	18.9	2007	15/08/2007	20:22:13	50.34	-177.57	ANDREANOF ISLANDS, ALEUTIAN IS.
6.7	53.8	2007	2/08/2007	3:21:47	51.36	-179.98	ANDREANOF ISLANDS, ALEUTIAN IS.
6.6	8.5	2006	8/07/2006	20:39:59	51.32	-179.27	ANDREANOF ISLANDS, ALEUTIAN IS.
6.6	53.9	2005	14/06/2005	17:10:15	51.16	-180.52	RAT ISLANDS, ALEUTIAN ISLANDS
6.7	13.2	2004	2/11/2004	10:02:11	49.21	-128.83	VANCOUVER ISLAND REGION
6.8	20	2004	28/06/2004	9:49:46	54.82	-134.49	QUEEN CHARLOTTE ISLANDS REGION
7.7	27.1	2003	17/11/2003	6:43:05	51.1	-181.36	RAT ISLANDS, ALEUTIAN ISLANDS
6.8	0.7	2003	23/06/2003	12:12:31	51.45	-183.29	RAT ISLANDS, ALEUTIAN ISLANDS
6.5	34.2	2003	15/06/2003	19:24:35	51.56	-183.15	RAT ISLANDS, ALEUTIAN ISLANDS
7.1	32.4	2003	17/03/2003	16:36:17	51.24	-182.09	RAT ISLANDS, ALEUTIAN ISLANDS
6.6	14.5	2003	19/02/2003	3:32:35	53.64	-164.73	UNIMAK ISLAND REGION
6.5	45.8	2002	7/11/2002	15:14:08	51.2	-180.63	RAT ISLANDS, ALEUTIAN ISLANDS

7	10.8	2002	3/11/2002	22:12:41	63.63	-147.61	CENTRAL ALASKA
6.7	19.7	2002	23/10/2002	11:27:20	63.55	-148.07	CENTRAL ALASKA
6.8	134.1	2001	28/07/2001	7:32:43	59.01	-155.09	SOUTHERN ALASKA
6.8	33	2001	10/01/2001	16:02:43	56.99	-153.46	KODIAK ISLAND REGION
6.6	42.2	2000	11/07/2000	1:32:26	57.45	-154.41	KODIAK ISLAND REGION
7	19.5	1999	6/12/1999	23:12:28	57.4	-154.57	KODIAK ISLAND REGION
6.8	25.8	1999	20/03/1999	10:47:45	51.56	-177.72	ANDREANOF ISLANDS, ALEUTIAN IS.
6.6	56.3	1999	28/01/1999	8:10:04	52.86	-169.17	FOX ISLANDS, ALEUTIAN ISLANDS
6.5	21.7	1997	17/12/1997	4:38:51	51.17	-181.17	RAT ISLANDS, ALEUTIAN ISLANDS
6.6	36	1997	26/03/1997	2:08:57	51.28	-180.47	RAT ISLANDS, ALEUTIAN ISLANDS
7.6	34.1	1996	10/06/1996	4:03:35	51.55	-177.61	ANDREANOF ISLANDS, ALEUTIAN IS.
6.8	4.1	1996	22/03/1996	3:24:17	51.28	-181.32	RAT ISLANDS, ALEUTIAN ISLANDS
6.8	12.3	1993	15/05/1993	21:52:22	51.34	-178.69	ANDREANOF ISLANDS, ALEUTIAN IS.
6.9	33.3	1993	13/05/1993	11:59:47	55	-160.39	ALASKA PENINSULA
6.9	13.6	1992	7/08/1992	18:19:20	57.59	-142.94	GULF OF ALASKA
6.7	10	1992	6/04/1992	13:54:40	50.65	-130.06	VANCOUVER ISLAND REGION
6.8	37.2	1991	30/05/1991	13:17:43	54.59	-161.59	ALASKA PENINSULA
6.5	20.2	1991	21/02/1991	2:35:34	58.43	-175.45	BERING SEA
6.5	100.2	1991	23/01/1991	1:12:28	51.96	-181.19	RAT ISLANDS, ALEUTIAN ISLANDS
6.6	209.7	1990	1/05/1990	16:12:21	58.85	-156.83	ALASKA PENINSULA
7	11	1989	4/09/1989	13:14:58	55.58	-156.86	SOUTH OF ALASKA
7.3	10	1988	6/03/1988	22:35:36	57.27	-142.79	GULF OF ALASKA
7.8	10	1987	30/11/1987	19:23:15	58.8	-142.6	GULF OF ALASKA
7.2	10	1987	17/11/1987	8:46:50	58.82	-143.25	GULF OF ALASKA
6.5	33.8	1987	21/06/1987	5:46:10	54.2	-162.66	ALASKA PENINSULA
6.8	25.9	1987	6/05/1987	4:06:15	51.26	-179.88	ANDREANOF ISLANDS, ALEUTIAN IS.
6.5	31	1986	12/09/1986	23:57:15	56.19	-153.4	KODIAK ISLAND REGION
6.8	17	1986	19/06/1986	9:09:10	56.39	-152.86	KODIAK ISLAND REGION
6.6	30	1985	9/10/1985	9:33:32	54.73	-159.65	SOUTH OF ALASKA
6.5	45.5	1983	14/02/1983	3:20:05	55.03	-159.19	ALASKA PENINSULA
6.5	28.8	1972	30/07/1972	21:45:15	56.77	-135.91	SOUTHEASTERN ALASKA

Table 2. Strong volcanic eruptions (VEI 3+), Alaska – Aleutian, from 1970 – 2013. Yellow highlight – VEI 4.

Spurr, Alsaka	1992	61.3	-152.25	4
Redoubt, Alaska	2009	60.48	-152.74	3
Redoubt, Alaska	1989	60.48	-152.74	3
Augustine, Alaska	2005	59.36	-153.43	3
Augustine, Alaska	1986	59.36	-153.43	4
Augustine, Alaska	1976	59.36	-153.43	4
Trident, Alaska	1974	58.24	-155.1	3
Ukinrek Maars, Alaska	1977	57.83	-156.51	3
Veniaminov, Alaska	1983	56.17	-159.38	3
Pavlof, Alaska	1986	55.42	-161.89	3
Pavlof, Alaska	1983	55.42	-161.89	3
Pavlof, Alaska	1981	55.42	-161.89	3
Pavlof, Alaska	1980	55.42	-161.89	3
Pavlof, Alaska	1980	55.42	-161.89	3
Pavlof, Alaska	1974	55.42	-161.89	3
Shishaldin, Aleutian	1995	54.76	-163.97	3
Shishaldin, Aleutian	1995	54.76	-163.97	3
Westdahl, Aleutian	1991	54.52	-164.65	3
Westdahl, Aleutian	1979	54.52	-164.65	3
Westdahl, Aleutian	1978	54.52	-164.65	3
Bogoslof, Aleutian	1992	53.93	-168.03	3
Okmok, Aleutian	2008	53.4	-168.17	4
Okmok, Aleutian	1997	53.4	-168.17	3
Okmok, Aleutian	1981	53.4	-168.17	3
Gareloi, Aleutian	1982	51.79	-178.79	3
Gareloi, Aleutian	1980	51.79	-178.79	3

The time-longitude plot of volcanic eruptions and earthquakes is shown in **Fig. 1**. An overall concentration of earthquake epicenters clearly shows the westward shift with time. The average propagation trend generated by the excel program indicates the speed of transmigration, 46 km/year or 130 m/day for earthquakes. A clear linear concentration from 2002 to 2013 is seen between the longitudes 170 and 182 W (= 178 E) in the western Aleutian. This linear trend may indicate eastward flow—its propagation speed being 50 km/year or 137 m/day.

Volcanic eruption plot, on the other hand, seems to indicate two trends; fast westward and slow eastward propagations. The former is 180-225 km/year or 500-620 m/day, and the latter 67 km/year or 185 m/day.

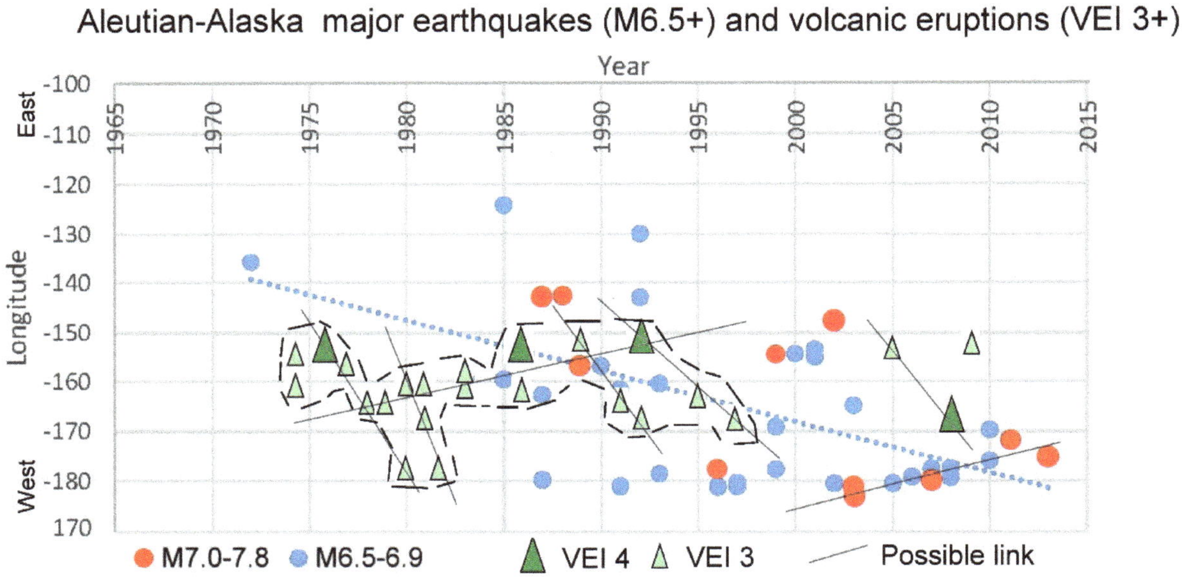

Figure 1. Longitude vs time plot of major volcanic eruptions and earthquakes. Overall trend shows the westward movement of volcanic and seismic energies. But also seen is the eastward propagation in both volcanic eruptions and earthquakes.

2) Alaska-Canada-Oregon

Major earthquakes with magnitude 6.0 or greater are listed in **Table 2**. The latitude-time plot of M6.0+ quakes in this region is shown in **Fig. 2**. The overall earthquake epicenter movement is southward. Its average rate of propagation is; 49 km/year or 135 m/day.

Table 2. List of M6.0+ earthquakes in Alaska, Canada and Oregon.

Magnitude	Depth (km)	Day	Time	Latitude	Longitude	Locality
6.1	37.4	12/07/1983	15:10:03	61.04	-147.37	SOUTHERN ALASKA
6.2	42.8	7/09/1983	19:22:04	60.97	-147.47	SOUTHERN ALASKA
6.4	16.9	28/02/1979	21:27:06	60.74	-141.55	SOUTHEASTERN ALASKA
6	14	28/06/1983	3:25:16	60.22	-141.27	SOUTHEASTERN ALASKA
7.2	10	17/11/1987	8:46:50	58.82	-143.25	GULF OF ALASKA
7.8	10	30/11/1987	19:23:15	58.8	-142.6	GULF OF ALASKA
6.1	1	6/01/2000	10:42:25	58.13	-136.93	SOUTHEASTERN ALASKA
6.3	9	12/11/2012	20:42:14	57.79	-142.86	GULF OF ALASKA
6.2	10	6/03/1988	23:14:36	57.74	-142.94	GULF OF ALASKA
6.9	13.6	7/08/1992	18:19:20	57.59	-142.94	GULF OF ALASKA
7.3	10	6/03/1988	22:35:36	57.27	-142.79	GULF OF ALASKA
6	21	16/01/1999	10:44:39	56.24	-147.42	GULF OF ALASKA
7.5	10	5/01/2013	8:58:19	55.39	-134.65	SOUTHEASTERN ALASKA
6.8	20	28/06/2004	9:49:46	54.82	-134.49	QUEEN CHARLOTTE ISLANDS REGION
6	20	12/07/2003	23:01:39	54.77	-134.35	QUEEN CHARLOTTE ISLANDS REGION
6.2	20	17/02/2001	20:11:30	53.97	-133.76	QUEEN CHARLOTTE ISLANDS REGION
7.8	14	28/10/2012	3:04:08	52.79	-132.1	QUEEN CHARLOTTE ISLANDS REGION
6	20	12/10/2001	5:02:35	52.71	-132.18	QUEEN CHARLOTTE ISLANDS REGION
6.3	9	28/10/2012	18:54:20	52.67	-132.6	QUEEN CHARLOTTE ISLANDS REGION
6.2	9	30/10/2012	2:49:02	52.37	-131.9	QUEEN CHARLOTTE ISLANDS REGION
6.6	8.7	17/11/2009	15:30:46	51.96	-131.6	QUEEN CHARLOTTE ISLANDS REGION
6	10	9/01/2008	14:40:01	51.64	-131.17	QUEEN CHARLOTTE ISLANDS REGION
6.6	10	5/01/2008	11:01:05	51.26	-130.76	QUEEN CHARLOTTE ISLANDS REGION
6.1	5.5	3/09/2013	20:19:06	51.23	-130.45	QUEEN CHARLOTTE ISLANDS REGION
6	9.9	4/09/2013	0:23:12	51.2	-129.9	QUEEN CHARLOTTE ISLANDS REGION
6.4	10	5/01/2008	11:44:48	51.15	-130.54	QUEEN CHARLOTTE ISLANDS REGION
6.7	10	6/04/1992	13:54:40	50.65	-130.06	VANCOUVER ISLAND REGION
6.6	11.4	24/04/2014	3:10:12	49.85	-127.44	VANCOUVER ISLAND REGION
6.4	25.5	19/07/2004	8:01:48	49.55	-127.01	VANCOUVER ISLAND REGION
6.5	22	9/09/2011	19:41:34	49.54	-126.89	VANCOUVER ISLAND REGION
6	10	2/07/1999	11:45:32	49.33	-129.15	VANCOUVER ISLAND REGION
6.1	13.7	8/11/2012	2:01:50	49.23	-128.48	VANCOUVER ISLAND REGION
6.7	13.2	2/11/2004	10:02:11	49.21	-128.83	VANCOUVER ISLAND REGION
6	10	11/01/2001	0:04:06	49.17	-128.9	VANCOUVER ISLAND REGION
6	10	6/10/1996	20:13:09	49.01	-127.89	VANCOUVER ISLAND REGION
6	10	14/09/2001	4:45:12	48.91	-128.26	VANCOUVER ISLAND REGION

6.7	56	28/02/2001	18:54:31	47.15	-122.63	WASHINGTON
6.2	10	2/06/2000	11:13:49	44.43	-130.18	OFF COAST OF OREGON
6.3	10	16/01/2003	0:53:15	44.21	-129.05	OFF COAST OF OREGON
6.3	10	10/01/2008	1:37:20	43.84	-126.91	OFF COAST OF OREGON
6.1	10	20/01/2000	9:41:53	43.7	-126.56	OFF COAST OF OREGON
6	8	11/04/2012	22:41:46	43.58	-127.64	OFF COAST OF OREGON
7.1	18	15/06/2005	2:50:55	41.45	-125.58	OFF COAST OF NORTHERN CALIFORNIA
6	10	21/02/2008	14:16:04	41.1	-114.88	NEVADA
6.6	12	17/06/2005	6:21:41	40.75	-126.47	OFF COAST OF NORTHERN CALIFORNIA
6.5	20.6	10/01/2010	0:27:41	40.67	-124.47	NEAR COAST OF NORTHERN CALIF.

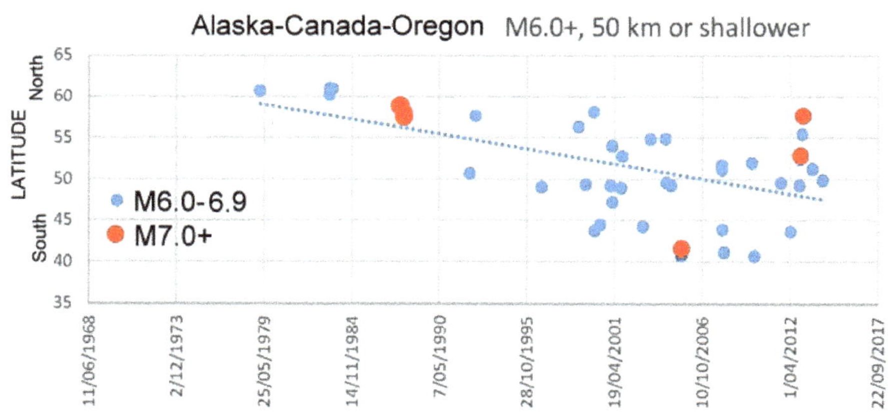

Figure 2. Time-latitude plot of m6.0+ quakes with depth 50 km or shallower.

3) Southern North America (Oregon-California-Mexico)

Major earthquakes in this region is listed below (**Table 3**). Their time-space graph is in **Fig. 3.**

Table 3. List of earthquakes, M6.0+, in Oregon, California and Mexico.

Magnitude	Depth (km)	Day	Time	Latitude	Longitude	Locality
6.4	0	26/03/1970	19:00:00	37.3	-116.53	SOUTHERN NEVADA

6.2	9.4	9/02/1971	14:00:40	34.4	-118.43	SOUTHERN CALIFORNIA
6.1	5	28/03/1975	2:31:06	42.04	-112.41	EASTERN IDAHO
6	11	26/11/1976	11:19:24	41.33	-125.66	OFF COAST OF NORTHERN CALIFORNIA
6.1	0	25/05/1980	16:33:44	37.6	-118.8	CALIFORNIA-NEVADA BORDER REGION
6.2	4.6	8/11/1980	10:27:31	41.15	-124.3	NEAR COAST OF NORTHERN CALIF.
6.5	10	3/11/1981	13:47:34	43.56	-127.7	OFF COAST OF OREGON
6.2	6.8	2/05/1983	23:42:37	36.24	-120.27	CENTRAL CALIFORNIA
6	9	22/07/1983	2:39:54	36.21	-120.37	CENTRAL CALIFORNIA
6.2	10	28/10/1983	14:06:07	44.1	-113.81	EASTERN IDAHO
6.1	5	10/09/1984	3:14:09	40.39	-126.8	OFF COAST OF NORTHERN CALIFORNIA
6.6	10	13/03/1985	19:34:57	43.5	-127.62	OFF COAST OF OREGON
6.6	10	18/06/1988	22:49:42	26.81	-111.04	GULF OF CALIFORNIA
6.9	7.6	18/10/1989	0:04:14	37.06	-121.79	CENTRAL CALIFORNIA
6.8	11	13/07/1991	2:50:14	42.19	-125.65	OFF COAST OF OREGON
6.4	7	16/08/1991	22:26:16	41.72	-125.42	OFF COAST OF NORTHERN CALIFORNIA
6.9	13.5	17/08/1991	22:17:15	41.85	-125.41	OFF COAST OF NORTHERN CALIFORNIA
6.2	12.4	23/04/1992	4:50:23	33.91	-116.48	SOUTHERN CALIFORNIA
7.1	15.1	25/04/1992	18:06:04	40.36	-124.05	NEAR COAST OF NORTHERN CALIF.
6.7	22.2	26/04/1992	11:18:26	40.47	-124.36	NEAR COAST OF NORTHERN CALIF.

7.2	1.1	28/06/1992	11:57:35	34.25	-116.48	SOUTHERN CALIFORNIA
6.3	10	28/06/1992	15:05:31	34.27	-116.78	SOUTHERN CALIFORNIA
6.1	25.9	17/05/1993	23:20:52	37.13	-117.81	CALIFORNIA-NEVADA BORDER REGION
6	10.7	21/09/1993	3:28:55	42.31	-122	OREGON
6.7	15.9	17/01/1994	12:30:54	34.14	-118.58	SOUTHERN CALIFORNIA
7.1	10	1/09/1994	15:15:53	40.44	-125.69	OFF COAST OF NORTHERN CALIFORNIA
6.1	14	12/09/1994	12:23:43	38.84	-119.65	CALIFORNIA-NEVADA BORDER REGION
6.4	10	19/02/1995	4:03:16	40.56	-125.53	OFF COAST OF NORTHERN CALIFORNIA
6	29.3	30/06/1995	11:59:00	24.75	-110.25	BAJA CALIFORNIA, MEXICO
6.6	12.2	28/08/1995	10:46:14	26.27	-110.36	GULF OF CALIFORNIA
7.1	0	16/10/1999	9:46:45	34.58	-116.44	SOUTHERN CALIFORNIA
6.1	17.7	13/11/2001	9:47:35	22.38	-106.95	NEAR COAST OF CENTRAL MEXICO
6.1	80	16/01/2002	23:10:19	22.5	-93.2	GULF OF MEXICO
6	120	30/01/2002	8:42:30	24.4	-95.6	GULF OF MEXICO
6.4	10	3/10/2002	16:08:29	23.3	-108.5	GULF OF CALIFORNIA
6.4	10.3	12/03/2003	23:41:33	26.65	-110.58	GULF OF CALIFORNIA
6.4	10	22/12/2003	19:15:56	35.67	-121.05	CENTRAL CALIFORNIA
6.5	14.1	4/01/2006	8:32:33	28.23	-112.13	GULF OF CALIFORNIA
6.1	9.8	1/09/2007	19:14:23	24.99	-109.66	GULF OF CALIFORNIA

6.9	7.8	3/08/2009	17:59:56	29.07	-112.91	GULF OF CALIFORNIA
6.2	10	3/08/2009	18:40:49	29.43	-113.76	GULF OF CALIFORNIA
7.2	5.2	4/04/2010	22:40:43	32.28	-115.26	CALIF.-BAJA CALIF. BORDER REGION
6.7	13.9	21/10/2010	17:53:13	24.79	-109.17	GULF OF CALIFORNIA
6.1	12	26/07/2011	17:44:20	25.1	-109.53	GULF OF CALIFORNIA
7	13	12/04/2012	7:15:48	28.7	-113.1	BAJA CALIFORNIA, MEXICO
6	9	12/04/2012	7:06:00	28.84	-113.03	BAJA CALIFORNIA, MEXICO
6.3	10	25/09/2012	23:45:24	24.67	-110.17	BAJA CALIFORNIA, MEXICO
6	14	8/10/2012	6:26:23	25.13	-109.57	GULF OF CALIFORNIA
6.4	13	14/12/2012	10:36:01	31.1	-119.66	OFF W. COAST OF BAJA CALIFORNIA
6.1	11.1	14/12/2012	10:36:18	32.41	-119.37	OFF COAST OF CALIFORNIA
6.4	1	19/10/2013	17:54:56	26.27	-110.18	GULF OF CALIFORNIA

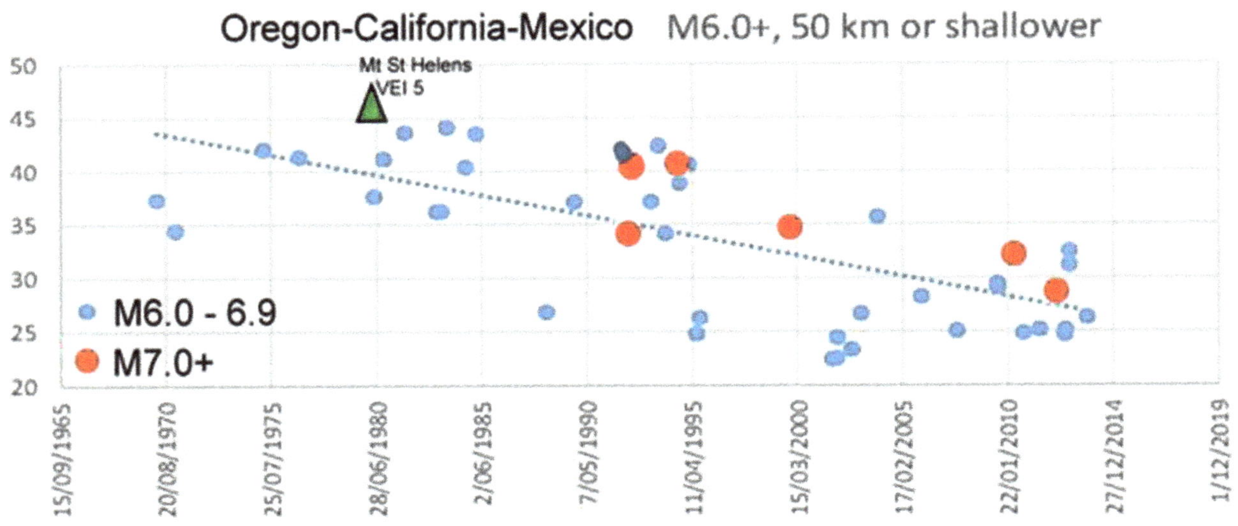

Figure 3. Time-latitude plot of M6.0+ earthquakes in Oregon-California-Mexico.

The time-latitude plot of the M6.0+ quakes (**Fig. 3**) indicates an average quake movement of 50 km/year

(= 137 m/day), which is similar to the Alaska-Canada-Oregon region and the Aleutians as well. A very powerful energy that caused the eruption of Mt. St Helens in May 1980 may have supplied energy for two very strong M7.0+ quakes down the stream in 1992 and 1994.

3. Discussion

1) Energy propagation speed in the Aleutians, Alaska, Canada, Western USA and Mexico.

A consistent trend has emerged by this study in regard to the lateral propagation rate of earthquake energy in all of the study areas; all fell in a relatively narrow window, 120–140 m/day. The westward flow trend recognized in the western Aleutians (137 m/day) also fell in the same range. The consistency in propagation rate in all study regions endorses the presence of the moving energy throughout the North America and the Aleutian Islands.

Whereas, volcanic energy propagation speed in the Aleutians is much faster than the earthquake speed especially for the westward volcanic energy flow which is almost four times faster. This can be due to the deep root of volcanoes, which generally ranges from 100 to 250 km depth (Blot, 1976 & 1981; Blot et al., 1974).

2) Eastward energy propagation in the Aleutians

As stated in foregoing pages there are clusters of quakes and volcanoes that suggest the eastward energy propagation in the Aleutian Islands. This trend is particularly strong from 1878 to 1992 in volcanic eruption pattern in the eastern Aleutian. From 2002 to 2013, an eastward moving earthquake clusters were developed in the eastern Aleutians.

We consider this eastward moving energy comes from Kamchatka, **Fig. 4**. This implies a complex system of flow channels in operation in the Aleutian region. A similar mixed flow pattern is also observed in the Caribbean region (Choi et al., in prep.).

3) Comparison of propagation speed with other areas.

In the deep section of Northern Chile, Choi (2014) showed the deep energy propagation from 214 km to 128 km depth was 410 m/day, and the 128 km to 35 km, 360 m/day. The major southward shallow (0–50 km) earthquake movement with time along the Pacific coast of South America was 250 m/day in average. This speed is much faster than that of the North America described above.

The Blot's energy transmigration speed for the Wadati-Benioff zone mainly derived from the western Pacific earthquakes is; 2.6 km/day at 600 km, 0.9 km/day at 200 km, and 0.5 km/day at 33 km (Grover, 1998). Interestingly the North American migration speed (120–140 m/day) is much slower than the western Pacific.

A flow speed obtained in the Caribbean by the study of Haiti quake (Choi, 2010) was 1.57 km/day for the depth range of 160 km to 7 km, much faster than the North American speed. This can be explained by the proximity to the energy source; Caribbean Sea is an oceanized mantle dome, where thermal energy rises directly from the deep mantle and the core (Choi, 2010).

Tsunoda et al. (2013) discussed the various earthquake energy transmigration speed and possible factors that affect the speed: proximity to the energy source, thermal gradient, geology, etc. The faster western Pacific flow speed than that of North America can be attributed to the proximity to the source, South Pacific superplume.

The propagation speed of strong volcanic eruptions in the Aleutians is; 500–620 m/day for the westward propagation and 185 m/day for the eastward propagation. The VE process propagation in Fuji Volcanic Zone was averaged to be about 400 m/day (Tsunoda et al., 2014).

4) Source of energy

The senior author (DC) presented at the IGC Brisbane 2012 the energy flow routes in the mantle (**Fig. 4**) based on the analysis of seismic tomography published by Kawakami et al. (1994), **Fig. 5**. Another more detailed global seismic tomographic images were prepared by Ohbayashi (unpublished), which was cited in Tsunoda (2010). The energy flow channels were constructed based on the distribution of slow mantle which is considered to contain gas or liquid which largely lowers down the p-wave propagation speed. The new data, which indicate westward flow in the Aleutians and the southeastward flow bifurcated at the Gulf of Alaska, may support the presence of the South Pacific-Hawaii-Gulf of Alaska energy flow channel in the middle mantle. However, further investigation on this energy flow route must be carried out from various from other viewpoints.

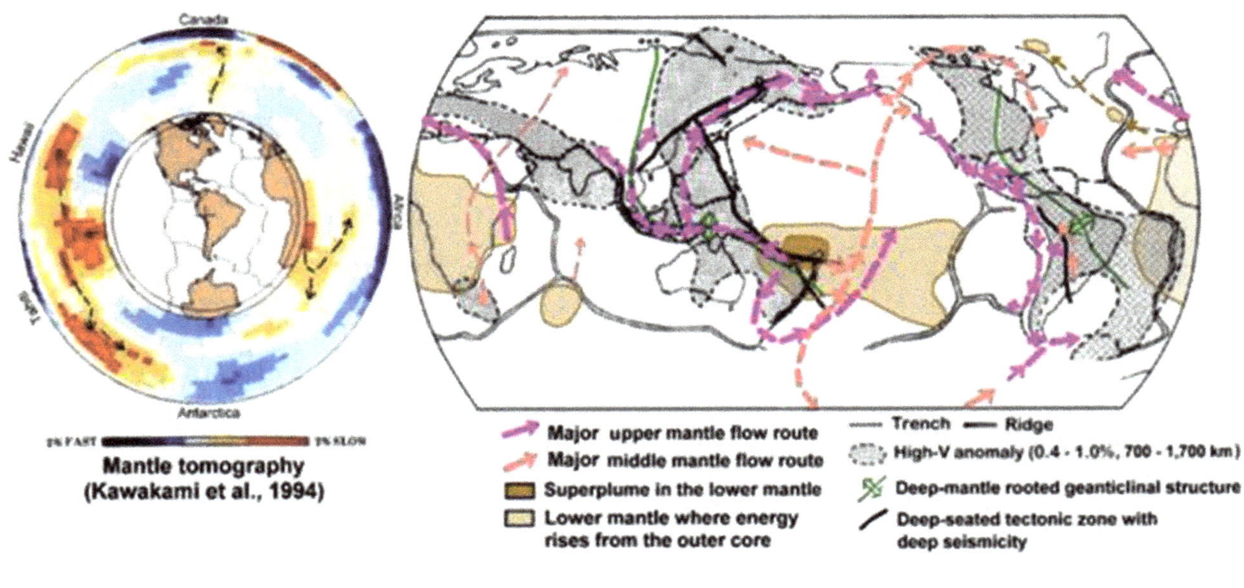

Figure 4. Mantle tomography by Kawakami et al. (1994), left, (broken lines added by one of authors, DC), and its interpretation, right (Choi, 2012, unpublished), showing the energy flow routes in the mantle. Energy rises from the superplume in the South Pacific and spread laterally in the middle mantel. The middle mantle flow crosses the Pacific Ocean under the Hawaii and emerges in the shallow mantle in the Gulf of Alaska, where it spreads into three branches, westward, southeastward and northward.

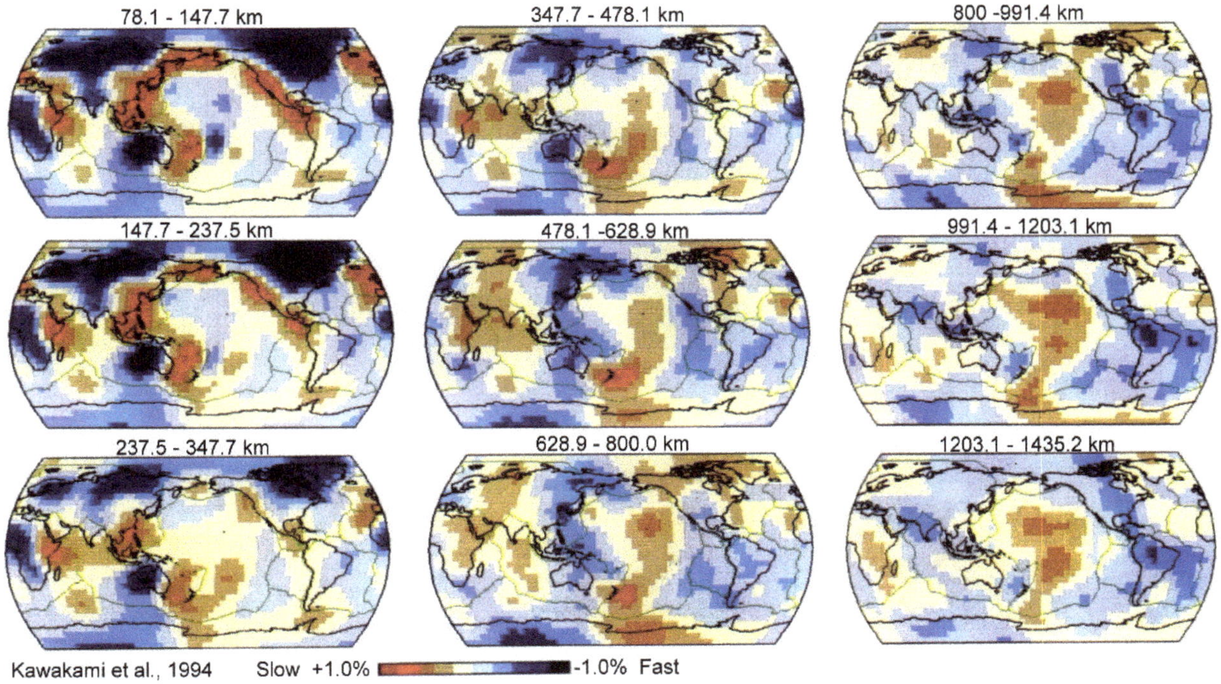

Figure 5. Seismic tomography by Kawakami et al. (1994). The distribution of low velocity mantle in the depth range from 1200 to 630 km runs from the SW Pacific to the Gulf of Alaska. This low velocity channel gradually retreats from 630 to 78 km depth. We consider this slow channel in the middle mantle is the flow route that supplies energy to the North America and the Aleutians. Also noted is a low velocity channel emanating from the Hawaii towards Japan in the depth range from 1200 km to 600 km.

The emergence of the middle mantle energy to the shallow depth in the Gulf of Alaska is corroborated by other data, **Fig. 6**, the latest sea surface temperature (SST) image for the last three-months from February to May, 2014. We noted that since the early December 2013 in the Gulf of Alaska region has become persistently high on a regional scale. The appearance of this regional high coincides with the rapid rise in seismic activities throughout the Pacific since 2010 with a rapid rise in March to April 2014 (**Fig. 7**; Choi et al., 2014).

The SST anomaly in **Fig. 6** displays another very intriguing fact. Readers will note a strong SST contrast between the northwestern Pacific (off Japan) and the northeastern Pacific (Gulf of Alaska); cold in the west and hot in the east. In addition, a sharp SST peak surrounded by low anomaly has appeared offshore Japan since November 2013 and lasted until early June 2014, for about six months. This phenomenon can be well explained by the sea-urchin model or joule heating proposed by Gregori (2013).

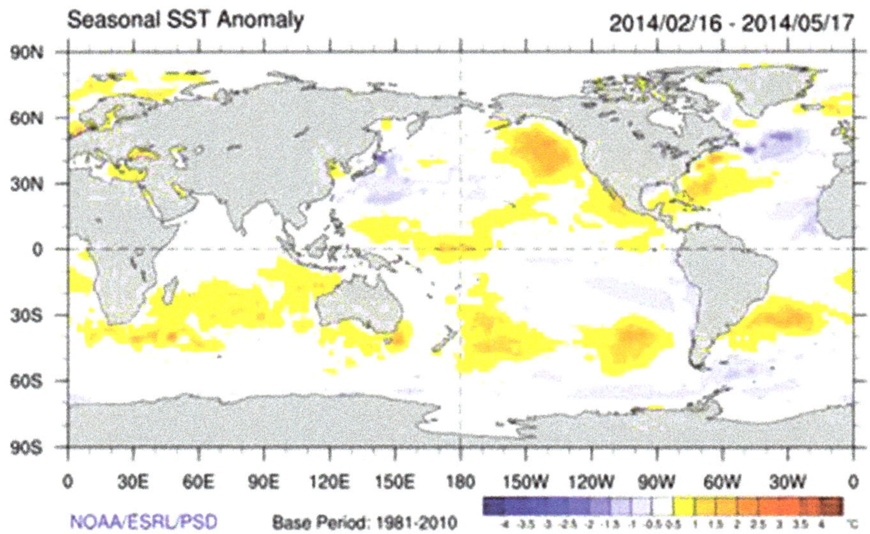

Figure 6. Sea surface Temperature anomaly image, from 16 February to 17 May 2014. Note an extremely high anomaly eye in the Gulf of Alaska region, which started appearing from December 2013 on. This SST image was taken from; http://www.esrl.noaa.gov/psd/map/clim/sst.shtml. Also note the cold patches off Japan.

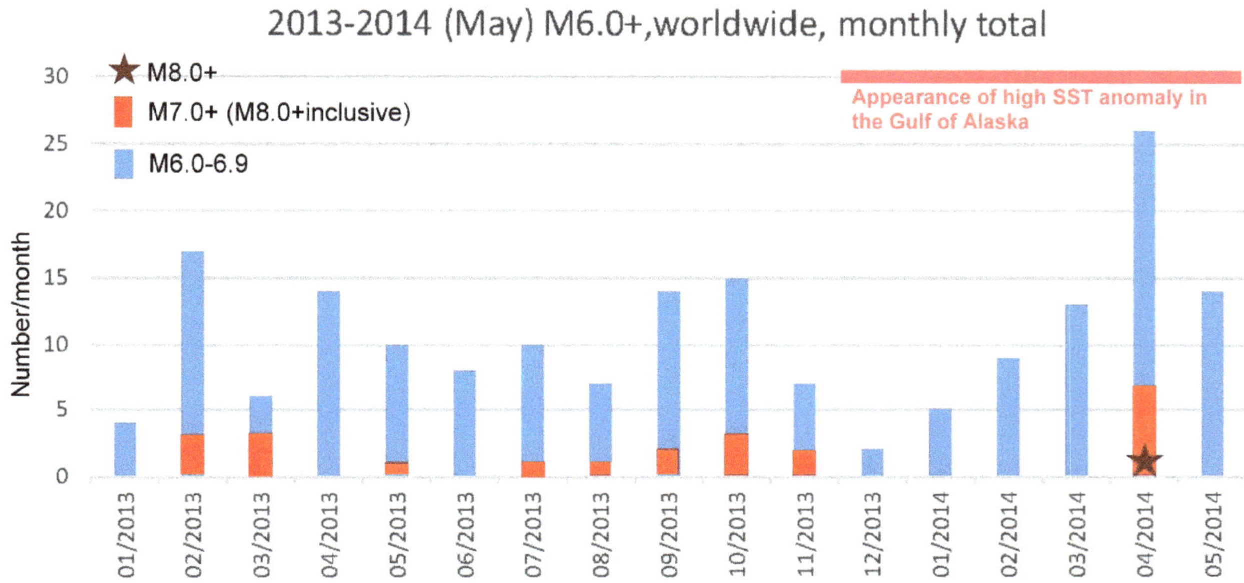

Figure 7. Monthly variation of the number of major earthquakes in the world. Most of the strong quakes in March to April 2014 occurred in the Pacific. Note the lowest number in December 2013 and a rapid rise towards April, 2014. This cycle coincides with the appearance of very strong SST high anomaly in the Gulf of Alaska (Fig. 6) since December 2013, which was countered by the cold patch off Japan.

4. Concluding remarks

This study clarified the energy flow occurring in North America and the Aleutians by analyzing the time-space distribution of major earthquakes and volcanic eruptions. Throughout the region, a consistent earthquake energy propagation rate in the shallow mantle and crust was obtained; 120–140 m/day. This rate is slower than that of the western Pacific and South America.

A complex flow pattern is observed in the Aleutian Islands—a mixture of eastward and westward flows: 1) Westward volcanic energy flow in the Aleutians is much faster than earthquakes, probably due to its deeper root, 100 to 250 km. Its speed is, 180–225 km/year or 500–620 m/day. 2) Eastward counter flow, which is slower than the westward flow, was also detected in the volcanic distribution pattern, whose average speed was calculated as 67 km/year or 185 m/day. The eastward flow of seismic energy is also observed from 1978 to 1990 in the eastern

Aleutian earthquake distribution; it turned out to be almost same in speed as the volcanic energy flow. This eastward flowing energy obviously came from the west, through the Japan-Kuril-Kamchatka route.

The source of energy in the study area is considered the South West Pacific. The energy transmigrates in the middle mantle channel through Hawaii, and upwells in the Gulf of Alaska, where the major portion of the energy bifurcates westward and southeastward along the coast of the Aleutian and North America.

The emergence of middle mantle energy to the shallow depth in the Gulf of Alaska is supported by the strong regional SST anomaly in the same region since December last year (2013). The discharged energy from this region should generate another cycle of strong earthquake and volcanic activities down the streams in the coming decades in the Aleutians and North America. Coupled with the deepening solar hibernation with the heightened endogenous energy release (Casey, 2012; Choi, 2013; Choi et al., 2014), it may bring a series of disastrous effects—catastrophic earthquakes and volcanic eruptions. Here, the knowledge of the upwelling site and propagation speed should help successfully predict the time and locality of these natural disasters.

The current work confirmed the veracity of energy migration inside the Earth through a complex nextwork of flow channels developed mainly in the Circum-Pacific Meso-Cenozoic mobile belts. Further studies are needed to fully understand the energy flow phenomena; flow channels, their geological control, Earth rotation effect, internal workings and processes of the deep Earth, and interaction with other planetary forces.

Acknowledgements: The authors thank John Casey, Bruce Leybourne and Giovanni Gregori for their comments and encouragement.

References cited

Blot, C., 1976. "Volcanisme et sismicité dans les arcs insulaires. Prévision de cesphénomènes." *Géophysique*, v. 13, Orstom, Paris, 206p.

Blot, C., 1981. "Deep root of andesitic volcanoes: new evidence of magma generation at depth in the Benioff zone." *Jour. Volcanology and Geothermal Research*, v. 10, p. 339–364.

Blot, C. and Choi, D. R., 2004. "Recent devastating earthquakes in Japan and Indonesia viewed from the seismic energy transmigration concept." *NCGT Newsletter*, no. 33, p. 3–12.

Blot, C., Defossez, M. and Monget, J. M., 1974. "Relations entre magmas volcaiqueset zones séismiquesdans les arcs Honshu et Hokkaido." *Bull. Volcanologique*, v. XXXVIII-4, p. 988–997.

Casey, J., 2012. *Cold Sun*. Trafford Publishing, 167p.

Choi, D. R., 2010. "The January 2010 Haiti seismic disaster viewed from the perspective of the energy transmigration concept and block tectonics." *NCGT Newsletter*, no. 54, p. 36–44.

Choi, D. R., 2011. "Geological analysis of the Great East Japan earthquake in March 2011." *NCGT Newsletter*, no. 59, p. 55–68.

Choi, D. R., 2013. "Earthquake/volcanic activities and solar cycles." *Global climate Status Report (GCSR)*, Edition 3-2013, p. 10–19.

Choi, D.R., 2014. "Seismo-electromagnetic energy flow observed in the 16 March 2014 M6.7 earthquake offshore Tarapaca, Chile." *NCGT Journal*, v. 2, no. 1, p. 61–65.

Choi, D. R., Casey, J., Maslov, L. and Tsunoda, F., 2014. "Increased seismic and magmatic activities since 1990 in the North and South Americas, and their relation to solar cycles." *Global Climate Status Report(GCSR)*, Edition 2-2014, p. 7–19.

Gregory, G. P., 2013. "Crustal stroms of continental/planetary scale." *NCGT Journal*, v. 1, no. 2, p. 40-64

Grover, J. C., 1998. "Volcanic eruptions and great earthquakes. Advanced warning techniques to master the deadly science." CopyRight Publishing Co. Pty Ltd., Brisbane. 272p.

Kawakami, s., Fujii, N. and Fukao, Y., 1994. "Frontiers of the earth and planetary science: a gallery of the planetary worlds." *Jour. Geol. Soc. Japan*, v. 100, p. I-VIII.

Meyheroff, A. A., Taner, I., Morris, A. E. L., Agocs, W .B, Kamen-Kaye, M., Bhat, M. I., Smoot, N. C. and Choi, D. R., (ed. Meyerhoff-Hull, D.), 1996. "Surge tectonics: a new hypothesis of global geodynamics." Kluwer Academic Publishers, 323p.

Tsunoda, F., 2011. "The March 2011 Great Offshore Tohoku-Pacific Earthquake from the perspective of the VE process." *NCGT Newsletter*, no. 59, p. 69–77.

Tsunoda, F., Choi, D.R. and Kawabe, T., 2013. "Thermal energy transmigration and fluctuation." *NCGT Journal*, v. 1, no. 2, p. 65–80.

Tsunoda, F., Kawabe, T. and Choi, D. R., 2014. "Transmigrating heat passing through the Aogashima Volcanic Island, Izu Volcanic Chain, Japan." *NCGT Journal*, v. 2, no. 2, p. 27-31.

Appendix 2, Item 11

Seismo-Volcanic Energy Propagation Trends in Central America and Their Relationship to Solar Cycles

(Published in the *New Concepts in Global Tectonics Journal* (*NCGT*), vol. 2,no. 3, Sep. 2014)

Dong R. Choi
International Earthquake and Volcano Prediction Center IEVPC), Canberra, Australia.
dchoi@ievpc.org

Abstract: Earthquake and volcanic energies in the Central America originate from the outer core under the Caribbean Sea and transmigrate to the Pacific coast through the oceanized horst structures. It is concluded that the direction of energy movement is controlled by the level of thermal energy input into the Caribbean dome from the outer core, which is reversely correlated to the solar cycle: During the solar declining cycle, earthquake and volcanic swarms moved northward, but during the rising cycle, southward.

Keywords: *earthquake, volcano, energy propagation, Caribbean, Central America, solar cycle*

Introduction

Our recent studies have clarified the seismo-volcanic energy flow patterns in many parts of the Circum Pacific regions (Blot, 1976; Meyerhoff et al., 1996; Tsunoda, 2009; Tsunoda et al., 2013 and 2014; Choi, 2014; Choi et al., 2014a and 2014b). These studies firmly established that the thermal energy that comes from the Earth's outer core to the surface spreads laterally along major deep fracture zones and mobile tectonic belts.

The Central America (**Fig. 1**) is unique in various ways: 1) The area is occupied by a series of oceanized seas including Caribbean Sea (Meso-Cenozoic) and Costa Rica Rift (Cenozoic), and possibly the Gulf of Mexico (Paleozoic oceanization? – Pratsch, 2010), 2) it is located on the axial area of the North-South American Geanticline (Choi, 2014), and 3) it is tectonically and magmatically very active in Cenozoic.

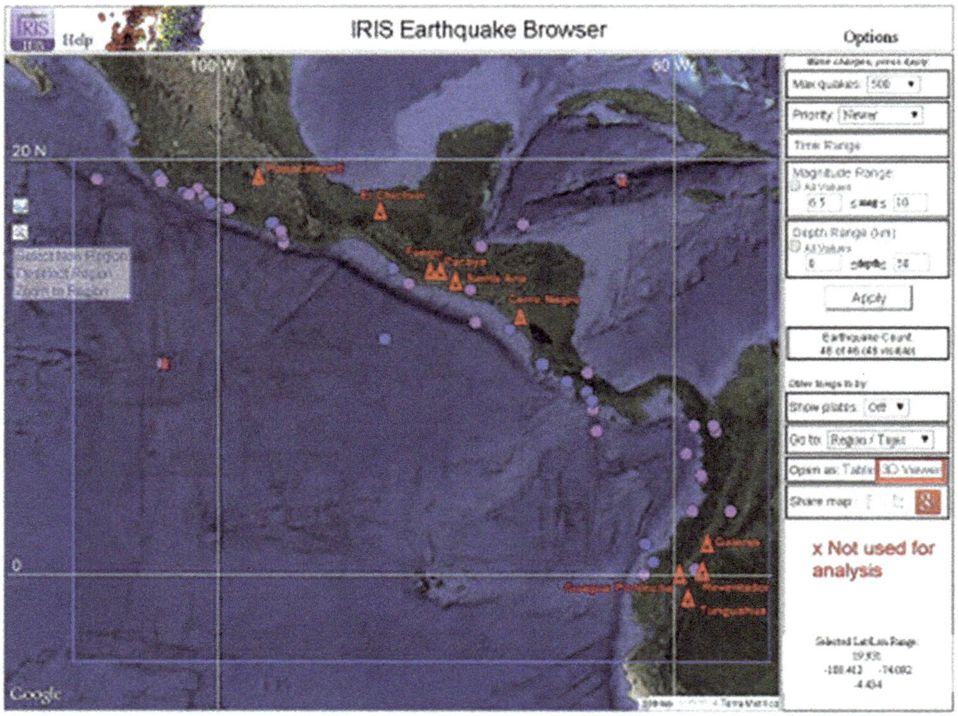

Figure 1. Earthquakes with magnitude 6.5 or greater, 50 km or shallower in depth, and occurred since 1970 in the Central America. Map generated by IRIS Seismic Monitor program (http://www.iris.edu/seismon/). Volcanoes indicated on this map erupted with the scale of VEI 3 or greater after 1970 and are listed in Table 2.

Furthermore, the present study clarified a complex energy propagation pattern with the deep-sourced energy coming from the east, the Caribbean Sea. The comparison of this pattern with solar cycle fluctuation revealed very interesting facts in regard to the energy movement. The author presents here some of the highlights of the study.

Earthquake propagation trends

Major earthquake (M6.0+) and volcanic eruption (VEI 3+) data were extracted from the IRIS (www.iris.edu/seismon/)/USGS NEIC (http://earthquake.usgs.gov/earthquakes/search), and USGS Volcanic Program websites (http://www.volcano.si.edu), respectively, and facilitated for the study (**Fig. 1**).

As seen in **Fig. 2** below, M6.0 to M6.9 quakes do not show particular trends, but the M6.5+ group quakes do show some significant patterns. The same fact was also observed in the South and North American earthquakes (Choi, 2014; Choi et al. 2014a). Based on this, in this study earthquakes with magnitude 6.5 or greater were used for study.

Other parameters of extracted quakes are; magnitude—6.5 or greater, depth range—50 km or shallower; and those occurred in the coastal area of the Central America within the latitudes between -4.5° and 20°. A list of earthquakes is shown in **Table 1**, and that of volcanic eruptions in **Table 2**: Their geographic positions are illustrated in **Fig. 1**.

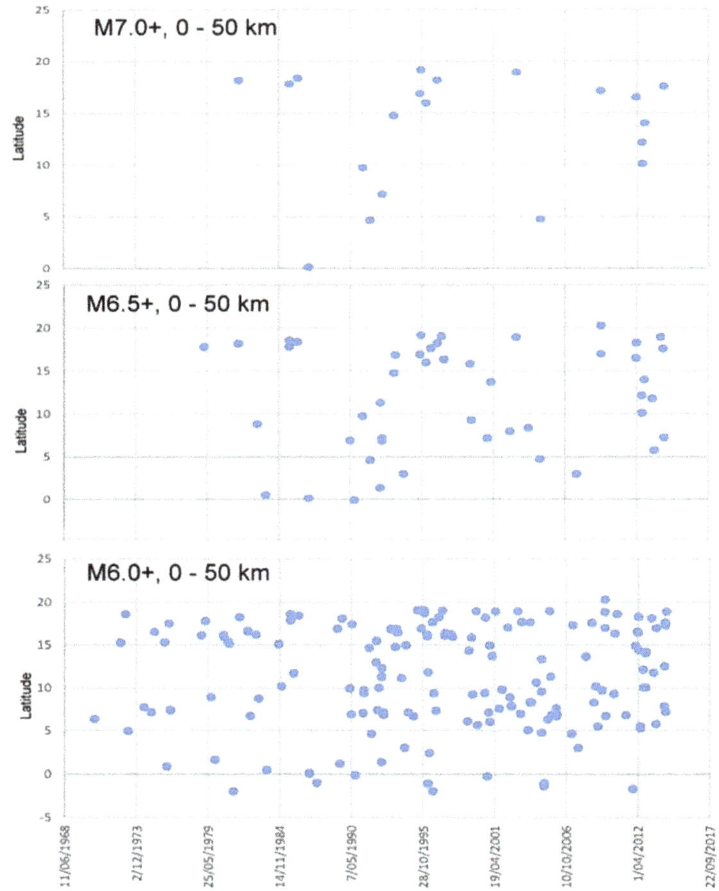

Figure 3. Latitude-time plots for comparison of various magnitude levels of strong earthquakes in Central America. No clear trend is seen in M6.0+ quakes (bottom), but trends appear in the M6.5+ quake group. Note a sudden appearance of numerous M6.5+ quakes from 1990 to 2004, and from 2012 to 2014. See also Figure 9.

Tale 1. List of very strong earthquakes (M6.5+) between the latitudes -0.5° and 20.00°, Central America.

Magni-tude	Depth (km)	Year	Date	Time	Latitude	Longitude	Locality
6.6	41.2	1974	8/10/1974	9:50:57	17.37	-61.99	LEEWARD ISLANDS
6.5	2.8	1979	14/03/1979	11:07:10	17.76	-101.3	NEAR COAST OF GUERRERO, MEXICO
7.3	28.1	1981	25/10/1981	3:22:16	18.18	-102.01	MICHOACAN, MEXICO
6.5	43.8	1983	3/04/1983	2:50:02	8.8	-83.11	COSTA RICA
6.6	38.8	1983	22/11/1983	14:21:01	0.48	-79.79	NEAR COAST OF ECUADOR
6.5	20.2	1985	16/03/1985	14:54:01	16.98	-62.46	LEEWARD ISLANDS
6.8	29.2	1985	19/09/1985	13:17:50	18.54	-102.32	MICHOACAN, MEXICO
7.6	42.1	1985	21/09/1985	1:37:15	17.81	-101.69	NEAR COAST OF GUERRERO, MEXICO
7	37.8	1986	30/04/1986	7:07:19	18.37	-103.01	NEAR COAST OF MICHOACAN, MEXICO
7	10	1987	6/03/1987	4:10:42	0.12	-77.8	COLOMBIA-ECUADOR BORDER REGION
6.5	9.6	1990	8/05/1990	0:01:40	6.9	-82.63	SOUTH OF PANAMA
7.3	4	1991	22/04/1991	21:56:51	9.7	-83.07	COSTA RICA
7.1	14.9	1991	19/11/1991	22:28:50	4.6	-77.41	NEAR WEST COAST OF COLOMBIA
6.6	50	1992	26/08/1992	9:43:11	1.36	-80.19	OFF COAST OF ECUADOR
6.7	50	1992	2/09/1992	1:51:04	11.29	-93.08	OFF COAST OF MEXICO
6.6	14.3	1992	17/10/1992	8:32:40	6.88	-76.76	NORTHERN COLOMBIA
7.2	10	1992	18/10/1992	15:11:59	7.15	-76.84	NORTHERN COLOMBIA
7.2	35.7	1993	10/09/1993	19:12:55	14.74	-92.69	NEAR COAST OF CHIAPAS, MEXICO
6.7	39	1993	24/10/1993	7:52:18	16.83	-98.73	NEAR COAST OF GUERRERO, MEXICO
6.7	32.6	1994	6/06/1994	20:47:43	2.99	-76.03	COLOMBIA
6.5	36.6	1995	19/01/1995	15:05:06	5.09	-72.94	COLOMBIA
7.4	44.4	1995	14/09/1995	14:04:34	16.88	-98.6	NEAR COAST OF GUERRERO, MEXICO
7.9	42.3	1995	9/10/1995	15:35:55	19.12	-104.2	NEAR COAST OF JALISCO, MEXICO
6.6	10	1995	1/12/1995	5:20:28	10.16	-104.02	NORTHERN EAST PACIFIC RISE
7.1	7.1	1996	25/02/1996	3:08:13	15.96	-98.09	OFF COAST OF GUERRERO, MEXICO
6.8	48.2	1996	15/07/1996	21:23:37	17.57	-101.05	NEAR COAST OF GUERRERO, MEXICO
7.2	49.3	1997	11/01/1997	20:28:28	18.17	-102.83	MICHOACAN, MEXICO
6.9	15	1997	1/05/1997	11:37:34	18.99	-107.27	OFF COAST OF JALISCO, MEXICO
6.9	3	1997	9/07/1997	19:24:10	10.5	-63.55	NEAR COAST OF VENEZUELA
6.9	31.4	1997	19/07/1997	14:22:08	16.34	-98.19	NEAR COAST OF GUERRERO, MEXICO
6.7	6.4	1999	11/07/1999	14:14:16	15.79	-88.32	HONDURAS
6.8	44.9	1999	20/08/1999	10:02:23	9.26	-84.06	COSTA RICA

6.5	17	2000	8/11/2000	6:59:59	7.14	-77.84	PANAMA-COLOMBIA BORDER REGION
6.5	7.6	2001	13/02/2001	14:22:05	13.7	-88.87	EL SALVADOR
6.5	10	2002	31/07/2002	0:16:44	7.94	-82.81	SOUTH OF PANAMA
6.5	33.4	2003	25/12/2003	7:11:11	8.36	-82.82	PANAMA-COSTA RICA BORDER REGION
7.2	15	2004	15/11/2004	9:06:55	4.74	-77.47	NEAR WEST COAST OF COLOMBIA
6.8	29	2007	10/09/2007	1:49:14	3	-77.9	NEAR WEST COAST OF COLOMBIA
7.3	29	2009	28/05/2009	8:24:48	16.81	-86.24	NORTH OF HONDURAS
6.7	10	2009	28/05/2009	8:24:57	20.28	-86.6	YUCATAN PENINSULA, MEXICO
7.4	20	2012	20/03/2012	18:02:47	16.49	-98.23	NEAR COAST OF GUERRERO, MEXICO
6.7	20	2012	11/04/2012	22:55:10	18.23	-102.69	MICHOACAN, MEXICO
7.4	28	2012	27/08/2012	4:37:19	12.14	-88.59	OFF COAST OF CENTRAL AMERICA
7.6	35	2012	5/09/2012	14:42:07	10.09	-85.32	COSTA RICA
7.3	24	2012	7/11/2012	16:35:46	13.99	-91.9	NEAR COAST OF GUATEMALA
6.5	37.4	2013	15/06/2013	17:34:28	11.79	-86.91	NEAR COAST OF NICARAGUA
6.7	12	2013	13/08/2013	15:43:15	5.77	-78.2	SOUTH OF PANAMA
6.5	28.5	2014	13/01/2014	4:01:04	19	-66.83	PUERTO RICO REGION
7.2	24	2014	18/04/2014	14:27:26	17.55	-100.82	GUERRERO, MEXICO

Table 2. List of volcanoes in Central America with VEI 3+ since 1970.

Name	**Year of major eruption**	**Latitude**	**Longitude**	**Eruption, VEI 3+**
Cerro Negra, Nicaragua	1971	12.51	-85.7	3
Fuego, Guatemala	1971	14.47	-90.88	3
Reventador, Ecuador	1973	-0.08	-77.66	3
Fuego, Guatemala	1974	14.47	-90.88	4
El Chichon, Mexico	1982	17.36	-93.23	5
Pacaya, Guatemala	1990	14.38	-90.6	3
Cerro Negra, Nicaragua	1992	12.51	-85.7	3
Popocatepetl, Mexico	1996	19.02	-98.62	3
Guagua Pichincha, Ecuador	1998	-0.17	-78.6	3
Tunguahua, Ecuador	1999	-1.47	-78.44	3

Reventador, Ecuador	2002	-0.08	-77.66	4
Galeras, Colombia	2004	1.22	-77.37	3
Pacaya, Guatemala	2004	14.38	-90.6	3
Santa Ana, El Salvador	2005	13.85	-89.63	3
Galeras, Colombia	2007	1.22	-77.37	3
Galeras, Colombia	2008	1.22	-77.37	3
Tunguahua, Ecuador	2010	-1.47	-78.44	3

As seen in **Fig. 4**, the Central America region displays a very interesting and complex trends; major northing trend from 1990 to 1997, a southing trend from 1988 to 2001, a brief northward movement from 2002 to 2003, and southward from 2004 to 2007. Another major southward trend is also present from 2009 to 2013.

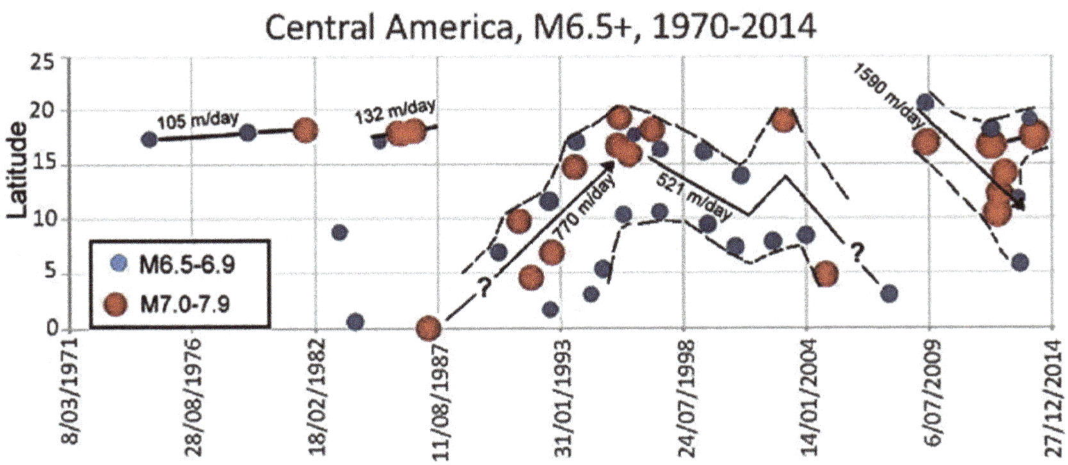

Figure 4. Time vs latitude plot of the M6.5+ earthquakes in the Central America. Lines indicate possible energy link.

Each group's average propagation speed was calculated by excel program (**Figs. 5 and 6** for example). The average speed of the major northward propagation (based on M7.0 quakes) from 1990 to 1997 is 280 km/year or 770 m/day. This group has numerous very strong quakes (M7.0 to 7.9). The major southward flow of the 1998–2001 group was calculated as 190 km/year or 521 m/day, much slower than the northward speed. Due to the paucity of samples, however, the speeds for the prior to 1990, and the 2002 to 2007 fluctuation cannot be given here. The latest southward group from 2009 to 2013 (**Fig. 6**) is much faster than others, calculated as 580 km/year or 1,590 m/day.

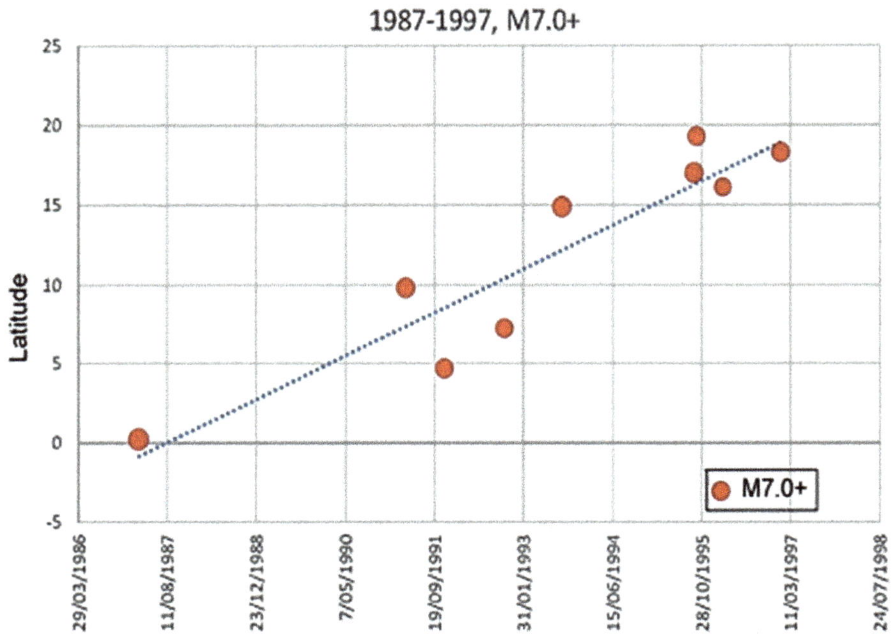

Figure 5. M7.0+ earthquakes from 1987 to 1997, northward propagation period. Average speed: 235 km/year or 644 m/day.

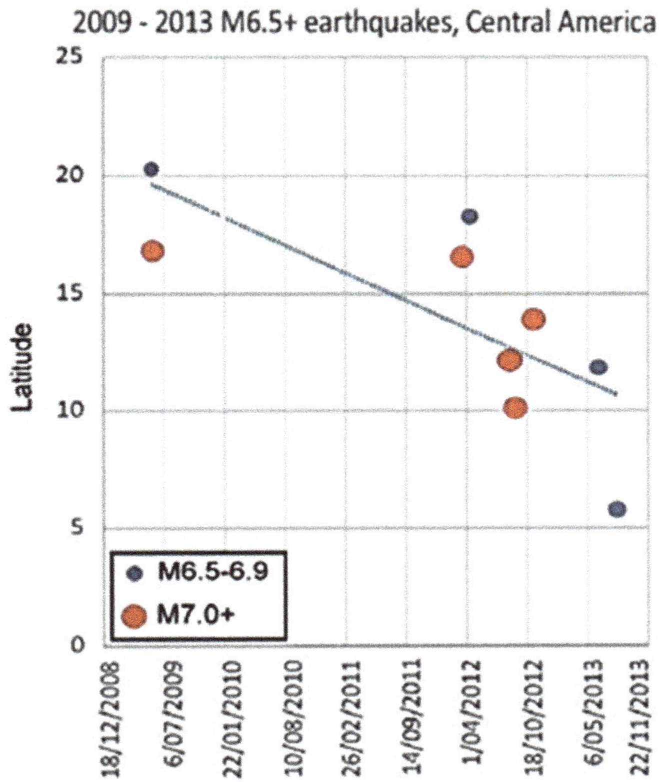

Figure 6. Time vs latitude plot of the M6.5+ earthquakes from 2009 to 2013 in the Central America with an average trend calculated by excel program (1,590 m/day). Note the absence of large quakes between 2010 and 2011 (rising period of solar cycle 24).

Propagation trend of volcanic eruptions

During the period of 1970 to 2013 a total of 17 major volcanic eruptions with VEI 3 or greater took place in the Central America (**Table 2; Figs. 1 & 6**) in two areas; north of 10° latitude and south of 2° latitude. No major eruptions (VEI3 or greater) are recorded in the Smithsonian Volcanic Program (http://volcano.si.edu/) during the study period in the Costa Rica, Panama and Colombia, despite the presence of numerous active volcanoes.

The largest eruption was El Chichon Volcano in Mexico with VEI 5 in 1982. The eruption occurred one year after the 1981 M7.3 Michoacan, Mexico quake, about 600 km west to the El Chichon Volcano. Other energy links are seen in 1990 to 1996 between eruptions and northward moving earthquakes. In this group, numerous very strong earthquakes and VEI 3 eruption coexisted in the same area.

Other outstanding features are very intensive volcanic activities in Ecuador to Colombia (near equator) from 1998 to 2010; first three years (1998-2002) in Ecuador, then, moving north to Columbia, triggered a series of VEI 3 Galera Volcano eruptions from 2004 to 2008.

Another possible energy link is seen prior to 1983 in the northward moving series of volcanoes extending in Nicaragua, Guatemala and Mexico (**Fig. 5**). It is noteworthy that its eruption scale increased with time. The energy propagation speed from Cerro Negro, Nicaragua in 1971 to El Chichon, Mexico in 1982 is 77 km/year or 102 m/day.

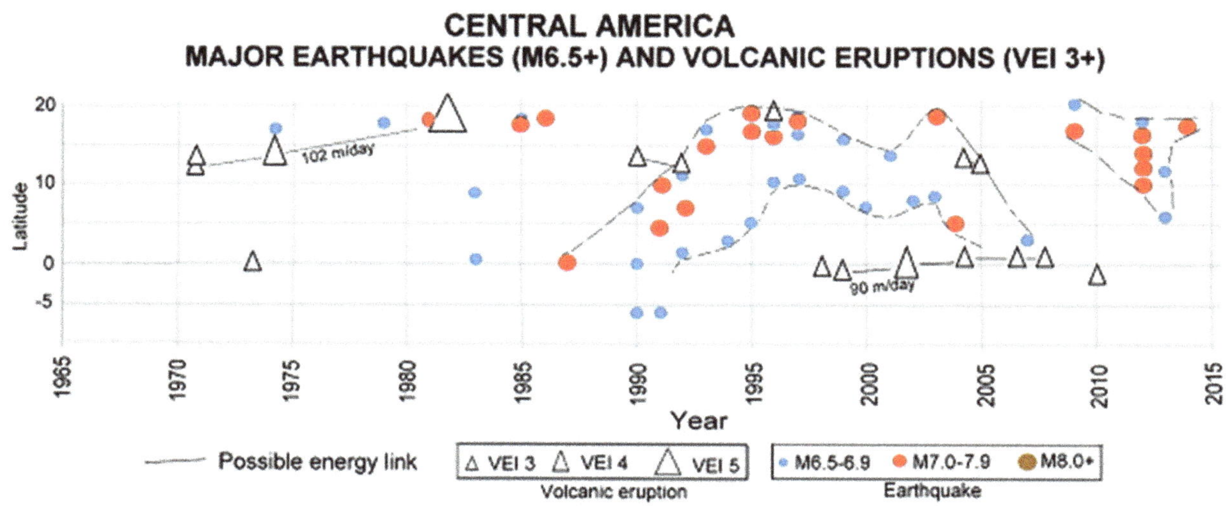

Figure 6. Year vs latitude plot of major earthquakes and volcanic eruptions, Central America from 1970 to 2014. Possible energy link of volcanoes and their propagation speed are indicated.

Discussion

Energy source of the Central American earthquakes and volcanoes

The author analysed the 2010 Haiti earthquake (Choi, 2010), in which he presented the oceanized Caribbean dome (**Fig. 7**). The paper contains detailed account of his idea and supporting data. The most important point that has to be emphasized here is the gigantic fracture system that is most likely connected with the lower mantle and possibly the outer core, as is well displayed in the mantle tomography by Karason and Van der Hilst (2000), **Fig. 7**.

The author already discussed the deep-seated fracture systems coinciding with the high-angle, narrow, fast mantle zones in tomographic images in many occasions (Choi, 2005 for example). Based on experiences in seismic data processing and interpretation for hydrocarbon exploration in the past, the author considers the p-wave velocity contrast in the tomographic images is strongly related to the amount of presence of fluid and/or gas in the mantle. The deep fractures are open systems through which outer core-derived materials rise to the Earth surface and finally escape to the atmosphere. Whereas, the slow mantle is primarily a closed system filled with fluid and gas. Interestingly the fast mantle zone under the Caribbean Sea is steeply inclined eastward at about 50 to 60 degrees from horizontal. Though obscured below 2,000 km, the continuation of the low velocity channel to the mantle bottom is almost beyond doubt if we take the relatively poor resolution in deep mantle into consideration.

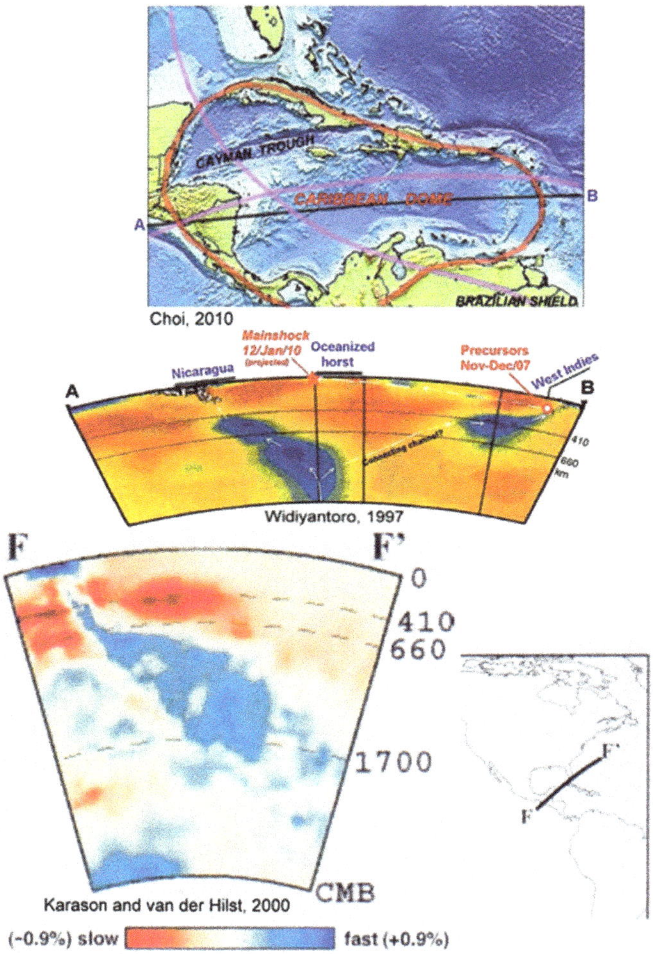

Figure 7. Seismo tomographic images of the Caribbean dome. Note inclined very distinctive deep mantle-rooted fast zone.

As discussed above, the seismic tomographic images convincingly suggests that the energy for the oceanization of the Caribbean dome was supplied from the outer core through the deep fracture systems under the Caribbean Sea. This energy supply is still continuing today to generate earthquakes and volcanic eruptions.

The Caribbean dome is situated on the gigantic Precambrian geanticlinal system that runs through North to South Americas and extends into the Rio Grand Ridge, off Brazil, Atlantic Ocean (Choi, 2013a and 2013b). Like the SE Asian region where the extensive oceanization took place since the late Mesozoic near the axis of the geanticlinal trend, Borneo-Vanuatu Geanticline (Choi, 2005 and 2013b), the Caribbean region has also been subject to the intensive oceanization.

Considering the structural highs which have been most extensively oceanized (Choi, 2010), the author speculates that the outer core energy, after reaching the upper mantle, flows mainly along the horst structures; the NE-SW horst running the middle of the Caribbean Sea from Hispaniola (Dominica) in the north to Panama in the south (which further runs into the Costa Rica Rift), and an E-W horst which forms Cayman Trough today.

IRIS Earthquake Browser

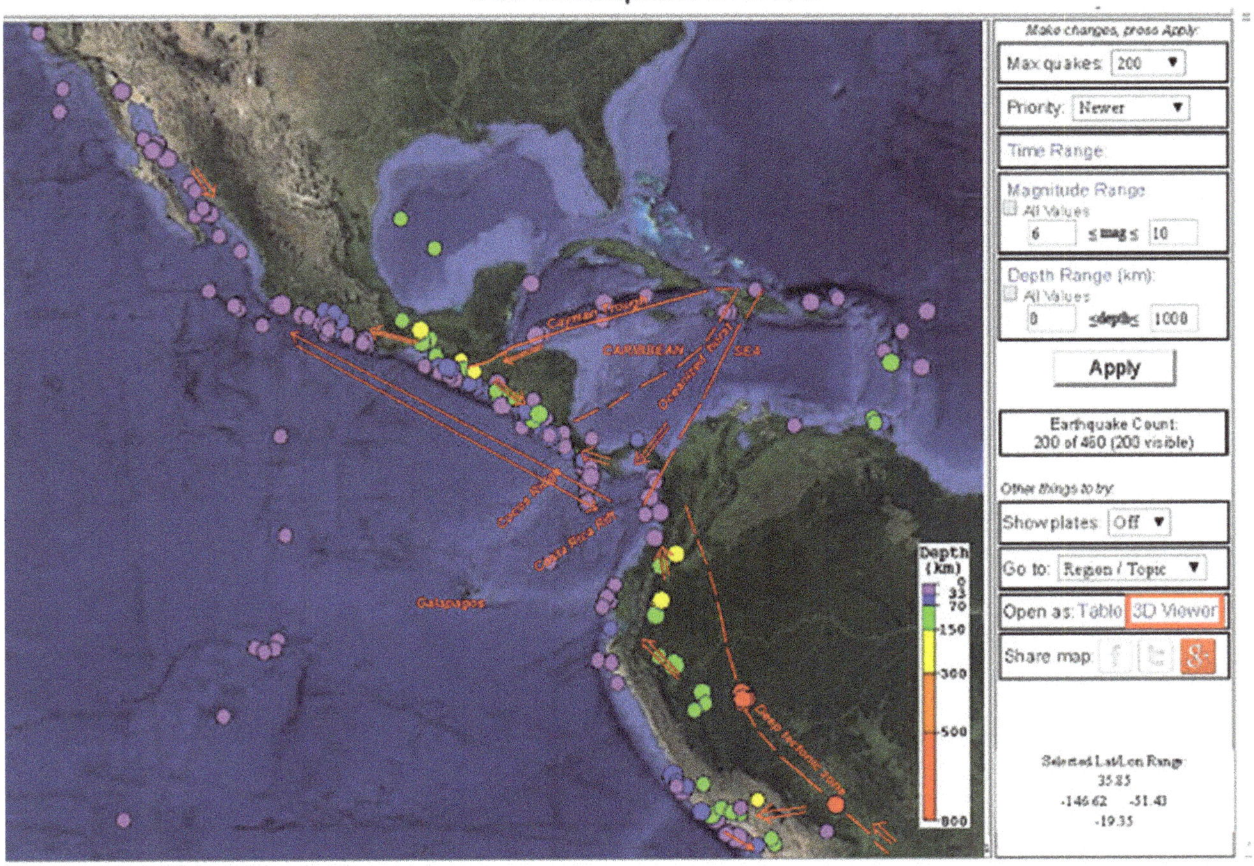

Figure 8. Suggested energy flow routes superimposed on the M6+ earthquake map (all depths, 1970–2014). For the deep tectonic zone in South America, see Choi, 2005.

Energy movement direction and solar cycle

The time-space distribution of the strong earthquakes (**Fig. 3**) and its comparison with solar cycle (**Fig. 9**) allow to reconstruct the following seismic history in the region. The story is especially intriguing in terms of reversed correlation between the solar cycle and earthquake frequency (Choi and Maslov, 2012; Choi et al., 2004a):

The deep-Earth sourced energy first arrived at Panama from the Caribbean in 1990. From there, with the onset of the active core phase, it started to move northward along the coast from 1990 while generated numerous strong quakes and volcanic eruptions on the way (start of sharp declining period of the Solar cycle 22), reached the Mexico region, where it stayed from 1995 to 1997 (solar cycle trough), and triggered powerful volcanic and seismic activities in the region. It then returned southward from 1997 to 2000 (Solar cycle 23 rising period–low energy period). The flow direction temporarily reversed and headed northward from 2003 to 2004 (Solar cycle 23 declining period), but again headed down south from 2004 to 2008 (trough between the solar cycles 23–24).

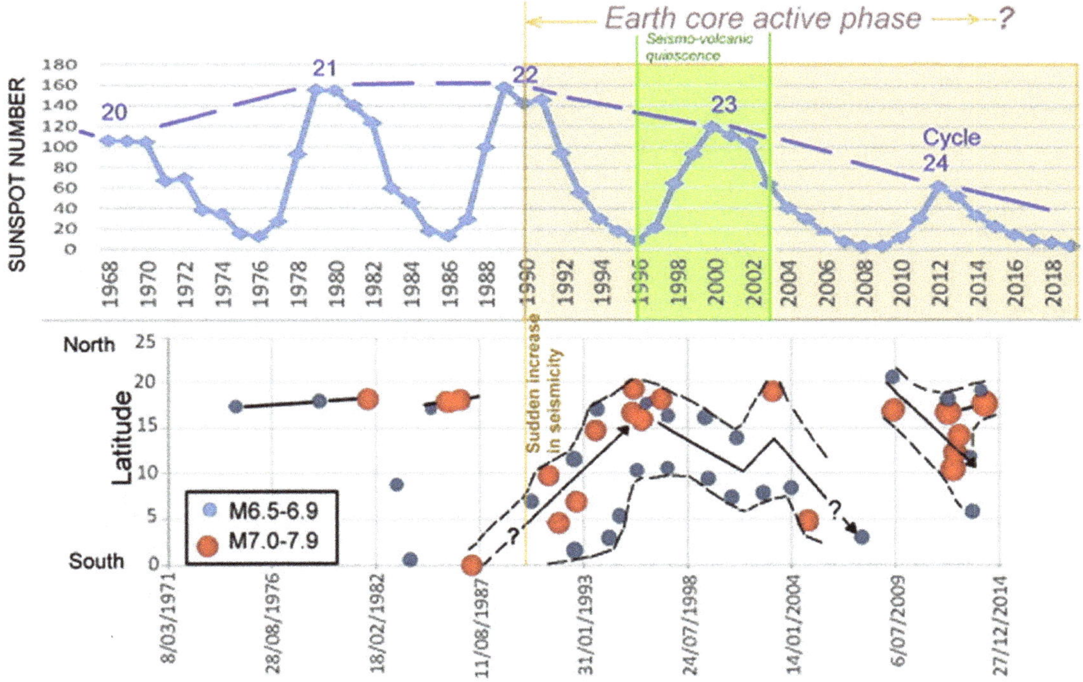

Figure 9. Solar cycles and earthquake propagation trend. Note a general trend that earthquakes move northward when the solar cycle is in decline, but southward when the solar cycle in rise, except for the period from 2005 to 2009 for which no data are available. A sudden increase in seismic activity from 1990 coincides with the start of declining period of a longer cycle obtained by tying the peaks of 11 year cycle. See Choi et al., 2014a. Seismo-volcanic quiescence cited from Choi (2010) and Tsunoda et al. (2013), and the Earth core active phase from Choi and Maslov (2010).

The above story testifies to an intricate relationship between the solar cycles and energy movement direction: During the solar cycle rise (relatively lower energy input from the Caribbean), seismo-volcanic energy moves south, but when declining (high energy phase), moves northward. This fact implies the thermal level (which is reversely related to the solar cycle) in the Caribbean Sea controls the direction of seismo-volcanic energy in the Pacific coast of the Central America.

Conclusions

This paper clarified a complex seismo-volcanic energy flow trend along the Pacific coast of the Central America, which is characterized by repeated cycle of northing and southing controlled by energy input fluctuation from the Caribbean Sea.

This pattern has not been seen in other areas in the Pacific Rim, such as Izu-Bonin Volcanic Islands Chain (Tsunoda et al., 2014), North America (Choi et al., 2014) and South America (Choi, 2014). This is considered to come from the unique geological setting of the Caribbean Sea: it forms a major dome structure formed by thermal energy directly derived from the outer core through gigantic deep fracture systems developed underneath. The Earth core activity is directly influencing the energy level fluctuation in the Caribbean Sea.

These facts have wide ramifications in understanding the Earth's geodynamic systems and internal workings that generates tectonic and magmatic activities at the Earth's surface.

References cited

Blot, C., 1976. "Volcanismeetsismicitédans les arcs insulaires. Prévision de cesphénomènes." *Géophysique*, v. 13, Orstom, Paris, 206p.

Choi, D. R., 2005. "Deep earthquakes and deep-seated tectonic zones: a new interpretation of the Wadati-Benioff zone." *Boll. Soc. Geol. It.*, vol. spec. no. 5, p. 79–118.

Choi, D. R., 2010. "The January 2010 Haiti seismic disaster viewed from the perspective of the energy transmigration concept and block tectonics." *NCGT Newsletter*, no. 54, p. 36–44.

Choi, D. R., 2010. "Global seismic synchronicity." *NCGT Newsletter*, no. 55, p. 66–73.

Choi, D. R., 2013a. "Continental rocks from the Rio Grande Ridge, South Atlantic." *NCGT Journal*, v. 1, no. 2, p. 2.

Choi, D. R., 2013b. "An Archean geanticline stretching from the South Pacific to Siberia." *NCGT Journal*, v. 1, no. 3, p. 45–55.

Choi, D. R., 2014. "Seismo-electromagnetic energy flow observed in the 16 March 2014 M6.7 earthquake offshore Tarapaca, Chile." *NCGT Journal*, v. 2, no. 1, p. 61–65.

Choi, D. R. and Maslov, L., 2012. "Earthquakes and solar activity cycles." *NCGT Newsletter*, no. 57, p. 85–97.

Choi, D. R., Casey, J., Maslov, L. and Tsunoda, F., 2014a. "Global increase in seismic and magmatic activities since 1990 and their relation to solar cycles." *The Global Climate Status Report (GCSR)*, Space and Science Research Corporation (SSRC), Orlando, USA, June 2014, p. 7–19.

Choi, D. R., Tsunoda, F. and Maslov, L., 2104b. "Seismo-volcanic energy propagation trends in the Aleutian Islands and North America." *NCGT Journal*, v. 2, no. 2, p. 7–12.

Karason, H. and Van der Hilst, R. D., 2001. "Improving global tomography models of P-wave speed I: incorporation of differential travel times for refracted and diffracted core phases" (PKP, Pdiff). *Jour. Gephys. Res*, v. 106, p. 6569–6587.

Meyheroff, A. A., Taner, I., Morris, A. E. L., Agocs, W. B, Kamen-Kaye, M., Bhat, M. I., Smoot, N. C. and Choi, D. R., (ed. Meyerhoff-Hull, D.). "Surge tectonics: a new hypothesis of global geodynamics." Kluwer Academic Publishers, 1996, 323p.

Pratsch, J-C., 2010. "Gulf of Mexico Basin – a collapsed Late Carboniferous mantle dome?" *NCGT Newsletter*, no. 55, p. 74–76.

Romanowicz, B., 2003. "Global mantle tomography: progress status in the past 10 years." *Annu. Rev. Earth Planet. Sci.*, v. 31, p. 303–328. (doi:10,1146/annurev.earth.31.091602.113555).

Tsunoda, F., 2009. "Habits of Earthquakes. Part 1: Mechanism of earthquakes and lateral thermal seismic energy transmigration." *NCGT Newsletter*, no.53, 38–46.

Tsunoda, F., Choi, D. R. and Kawabe, T., 2013. "Thermal energy transmigration and fluctuation." *NCGT Journal*, v. 1, no. 2, p. 65–80.

Tsunoda, F., Kawabe, T. and Choi, D. R., 2014. "Transmigrating heat passing through the

Aogashima Volcanic Island, Izu Volcanic Chain, Japan." *NCGT Journal*, v. 2, no. 2, p. 27–31.

Widiyantoro, S., 1997. "Studies of seismic tomography on regional and global scale." Ph.D. Thesis, Australia National University, Canberra, Australia.

Appendix 3

Most Destructive Earthquakes

Earthquakes with 50,000 or More Deaths[9]

*Listed in order of greatest number of deaths

[9] . Source: Wikipedia, the Free Encyclopaedia

Date UTC	Location	Deaths	Magnitude	Comments
01/23/1556	Shaanxi (Shensi), China	830000	8	The earthquake occurred near Huaxian, Shaanxi (formerly Shensi), China, about 50 miles (80 km) east-northeast of Xi'an, the capital of Shaanxi. More than 830,000 people were killed. Damage extended as far away as Taiyuan, the capital of Shanxi (formerly Shansi) and about 270 miles (430 km) northeast of the epicenter. There are felt reports as far away as Liuyang in Hunan, more than 500 miles (800 km) away. Geological effects reported with this earthquake included ground fissures, uplift, subsidence, sand blows, liquefaction and landslides. Most towns in the damage area reported city walls collapsed, most to all houses collapsed, and many of the towns reported ground fissures with water gushing out (i.e., liquefaction and sand blows). Gu et al. say that "the identified death toll of soldiers and civilians was 830,000, and the unidentified was uncountable." The earthquake was felt in all or parts of 9 provinces: Anhui, Gansu, Hebei, Hubei, Henan, Hunan, Shaanxi, Shandong, and Shanxi. The maximum intensity is XI in the Huaxian-Weinan area and the estimated magnitude is 8. Additional details from Gu et al.: In Huaxian, "city walls, temples, offices and civilian houses were demolished, without a single wall left standing . . . The ground fissured and sunk. Water gushed out and formed canals. Sixty percent of the people (several tens of thousands were killed or injured" In Weinan (15 miles [24 km] west of Huaxian), "city walls, temples, storehouses, offices and civilian houses collapsed totally . . . In the city, the ground sunk for more than 3 meters. Fifty percent of the people were killed." In Xi'an [one of China's major cities then as it is now], "city walls, storeyed buildings and terraces collapsed. Most temples were destroyed. More than half of the houses toppled down. Only 10–20% of the walls were left standing. The ground fissured crisscross. Thirty percent of the people were killed. Even as far away as Taiyuan, "houses were destroyed in great numbers." In many references, this earthquake is referred to as the "Shensi Province earthquake of 1556" using the old spelling for the province. [310]

Date UTC	Location	Deaths	Magnitude	Comments
01/12/2010	Haiti region	316000	7.0	According to official estimates, 316,000 people were killed, 300,000 injured, 1.3 million displaced, 97,294 houses destroyed, and 188,383 damaged in the Port-au-Prince area and in much of southern Haiti. Other estimates suggest substantially lower numbers of casualties, perhaps as low as fewer than 100,000. The casualties include at least 4 people killed by a local tsunami in the Petit Paradis area near Leogane. Tsunami waves were also reported at Jacmel, Les Cayes, Petit Goave, Leogane, Luly and Anse a Galets. The tsunami had recorded wave heights (peak-to-trough) of 12 cm at Santo Domingo, Dominican Republic, and 2 cm at Christiansted, US Virgin Islands. Uplift was observed along the coast from Leogane to L'Acul, and subsidence was observed along the coast from Grand Trou to Port Royal. Felt (VIII) at Leogane; (VII) at Carrefour, Port-au-Prince and Petionville; (VI) at Vieux Bourg d'Aquin; (V) at Port-de-Paix. Felt (V) at La Vega, Moca and San Cristobal; (IV) at Bani, Bonao, Luperon, Nagua, Puerto Plata, Santiago,Santo Domingo and Sosua, Dominican Republic. Felt throughout Haiti and the Dominican Republic. Felt (III) at Oranjestad, Aruba; (IV) at Santiago de Cuba and (III) at Guantanamo, Cuba; (II) in the Kingston-Mona area, Jamaica; (II) at Carolina and San Juan, Puerto Rico; (III) at Cockburn Harbour and (II) at Cockburn Town, Turks and Caicos Islands; (III) at Maracaibo and (II) at Caracas, Venezuela. Felt in parts of The Bahamas, Puerto Rico, and the US Virgin Islands and as far as southern Florida, northern Colombia and northwestern Venezuela.
07/27/1976	Tangshan, China	242769	7.5	Official casualty figure is 242,769 deaths. Estimated death toll as high as 655,000. 799,000 injured and extensive damage in the Tangshan area. Damage extended as far as Beijing.
08/09/1138	Syria, Aleppo	230000		

Date UTC	Location	Deaths	Magnitude	Comments
12/26/2004	Sumatra	227898	9.1	This is the third largest earthquake in the world since 1900 and is the largest since the 1964 Prince William Sound, Alaska, earthquake. In total, 227,898 people were killed or were missing and presumed dead and about 1.7 million people were displaced by the earthquake and subsequent tsunami in 14 countries in South Asia and East Africa. (In January 2005, the death toll was 286,000. In April 2005, Indonesia reduced its estimate for the number missing by over 50,000.) The earthquake was felt (IX) at Banda Aceh, (VIII) at Meulaboh and (IV) at Medan, Sumatra and (III-V) in parts of Bangladesh, India, Malaysia, Maldives, Myanmar, Singapore, Sri Lanka and Thailand. The tsunami caused more casualties than any other in recorded history and was recorded nearly worldwide on tide gauges in the Indian, Pacific, and Atlantic Oceans. Seiches were observed in India and the United States. Subsidence and landslides were observed in Sumatra. A mud volcano near Baratang, Andaman Islands, became active on December 28 and gas emissions were reported in Arakan, Myanmar.
12/22/856	Iran, Damghan	200000		
12/16/1920	Haiyuan, Ningxia (Ning-hsia), China	200000	7.8	Total destruction (XII—the maximum intensity on the Mercalli scale) in the Lijunbu-Haiyuan-Ganyanchi area. Over 73,000 people were killed in Haiyuan County. A landslide buried the village of Sujiahe in Xiji County. More than 30,000 people were killed in Guyuan County. Nearly all the houses collapsed in the cities of Longde and Huining. Damage (VI-X) occurred in 7 provinces and regions, including the major cities of Lanzhou, Taiyuan, Xi'an, Xining and Yinchuan. It was felt from the Yellow Sea to Qinghai (Tsinghai) Province and from Nei Mongol (Inner Mongolia) south to central Sichuan (Szechwan) Province. About 200 km (125 mi) of surface faulting was seen from Lijunbu through Ganyanchi to Jingtai. There were large numbers of landslides and ground cracks throughout the epicentral area. Some rivers were dammed; others changed course. Seiches from this earthquake were observed in 2 lakes and 3 fjords in western Norway. Although usually called the Kansu (now Gansu) earthquake by Western sources, the epicenter and highest intensities are clearly within Ningxia Autonomous Region. [310,92,316]
03/23/893	Iran, Ardabil	150000		

Date UTC	Location	Deaths	Magnitude	Comments
09/01/1923	Kanto (Kwanto), Japan	142800	7.9	Extreme destruction in the Tokyo -Yokohama area from the earthquake and subsequent firestorms, which burned about 381,000 of the more than 694,000 houses that were partially or completely destroyed. Although often known as the Great Tokyo Earthquake (or the Great Tokyo Fire), the damage was apparently most severe at Yokohama. Damage also occurred on the Boso and Izu Peninsulas and on O-shima. Nearly 2 m (6 ft) of permanent uplift was observed on the north shore of Sagami Bay and horizontal displacements of as much as 4.5 m (15 ft) were measured on the Boso Peninsula. A tsunami was generated in Sagami Bay with wave heights as high as 12 m (39 ft) on O-shima and 6 m (20 ft) on the Izu and Boso Peninsulas. Sand blows were noted at Hojo which intermittently shot fountains of water to a height of 3 m (10 ft). [303,6,312,321]
10/05/1948	Ashgabat (Ashkhabad), Turkmenistan (Turkmeniya, USSR)	110000	7.3	Extreme damage in Ashgabat (Ashkhabad) and nearby villages, where almost all brick buildings collapsed, concrete structures were heavily damaged, and freight trains were derailed. Damage and casualties also occurred in the DarrehGaz area, Iran. Surface rupture was observed both northwest and southeast of Ashgabat. Many sources list the casualty total at 10,000, but a news release on December 9, 1988 advised that the correct death toll was 110,000. [233,191]
09/27/1290	China, Chihli	100000		

Date UTC	Location	Deaths	Magnitude	Comments
05/12/2008	Eastern Sichuan, China	87587	7.9	At least 69,195 people killed, 374,177 injured and 18,392 missing and presumed dead in the Chengdu-Lixian-Guangyuan area. More than 45.5 million people in 10 provinces and regions were affected. At least 15 million people were evacuated from their homes and more than 5 million were left homeless. An estimated 5.36 million buildings collapsed, and more than 21 million buildings were damaged in Sichuan and in parts of Chongqing, Gansu, Hubei, Shaanxi and Yunnan. The total economic loss was estimated at 86 billion U.S. dollars. Beichuan, Dujiangyan, Wuolong and Yingxiu were almost completely destroyed. Landslides and rockfalls damaged or destroyed several mountain roads and railways and buried buildings in the Beichuan-Wenchuan area, cutting off access to the region for several days. At least 700 people were buried by a landslide at Qingchuan. Landslides also dammed several rivers, creating 34 barrier lakes which threatened about 700,000 people downstream. A train was buried by a landslide near Longnan, Gansu. At least 2,473 dams sustained some damage, and more than 53,000 km of roads and 48,000 km of tap water pipelines were damaged. About 1.5 km of surface faulting was observed near Qingchuan, surface cracks and fractures occurred on three mountains in the area, and subsidence and street cracks were observed in the city itself. Maximum intensity XI was assigned in the Wenchuan area. Felt (VIII) at Deyang and Mianyang; (VII) at Chengdu; (VI) at Luzhou and Xi'an; (V) at Chongqing, Guozhen, Lanzhou, Leshan, Wu'an, Xichang and Ya'an. Felt in much of central, eastern, and southern China, including Beijing, Guangzhou, Hefei, Nanjing, Shanghai, Tianjin, Wuhan, and in Hong Kong. Also felt in parts of Bangladesh, Taiwan, Thailand, and Vietnam. Seiches were observed at Kotalipara, Bangladesh.

Date UTC	Location	Deaths	Magnitude	Comments
10/08/2005	Pakistan	86000	7.6	At least 86,000 people killed, more than 69,000 injured, and extensive damage in northern Pakistan. The heaviest damage occurred in the Muzaffarabad area, Kashmir, where entire villages were destroyed and at Uri where 80 percent of the town was destroyed. At least 32,335 buildings collapsed in Anantnag, Baramula, Jammu, and Srinagar, Kashmir. Buildings collapsed in Abbottabad, Gujranwala, Gujrat, Islamabad, Lahore and Rawalpindi, Pakistan. Maximum intensity VIII. Felt (VII) at Topi; (VI) at Islamabad, Peshawar and Rawalpindi; (V) at Faisalabad and Lahore. Felt at Chakwal, Jhang, Sargodha and as far as Quetta. At least 1,350 people killed and 6,266 injured in India. Felt (V) at Chandigarh and New Delhi; (IV) at Delhi and Gurgaon, India. Felt in Gujarat, Haryana, Himachal Pradesh, Madhya Pradesh, Punjab, Rajasthan, Uttaranchal, and Uttar Pradesh, India. At least one person killed and some buildings collapsed in Afghanistan. Felt (IV) at Kabul and (III) at Bagrami, Afghanistan. Felt (III) at Kashi, China and (II) at Dushanbe, Tajikistan. Also felt at Almaty, Kazakhstan. An estimated 4 million people in the area were left homeless. Landslides and rockfalls damaged or destroyed several mountain roads and highways cutting off access to the region for several days. Landslides also occurred farther north near the towns of Gilgit and Skardu, Kashmir. Liquefaction and sandblows occurred in the western part of the Vale of Kashmir and near Jammu. Landslides and rockfalls also occurred in parts of Himachal Pradesh, India. Seiches were observed in Haryana, Uttar Pradesh and West Bengal, India, and in many places in Bangladesh.
11/1667	Caucasia, Shemakha	80000		
11/18/1727	Iran, Tabriz	77000		

Date UTC	Location	Deaths	Magnitude	Comments
12/28/1908	Messina, Italy	72000	7.2	Over 40% of the population of Messina and more than 25% of Reggio di Calabria killed by the earthquake and tsunami, as well as by fires in some parts of Messina. Casualty toll is based on census data of 1901–1911; some estimates are as high as 110,000. Severe damage in large parts of Calabria and Sicily. Felt throughout Sicily and north to Naples and Campobasso. Also felt on Malta, in Montenegro, and Albania and on the Ionian Islands. Tsunami heights of 6-12 m (20–39 ft) observed on the coast of Sicily south of Messina and heights of 6–10 m (20–33 ft) observed along the coast of Calabria. Aftershocks continued into 1913. [301,299,A-75]
05/31/1970	Chimbote, Peru	70000	7.9	About 50,000 people were killed, 20,000 missing and presumed dead, and 150,000 injured in Ancash and La Libertad Departments from the earthquake and a catastrophic debris avalanche of rock, ice, and mud which buried the town of Yungay, which had a population of about 20,000.
11/01/1755	Portugal, Lisbon	70000	8.7	This earthquake occurred on All Saint's Day while many of the 250,000 inhabitants of Lisbon were in church. Stone buildings swayed violently and then collapsed on the population. Many who sought safety on the riverfront were drowned by a large tsunami. Fire ravaged the city. One quarter of Lisbon's population perished. This earthquake had a profound effect on the intellectual outlook of Europe.
01/11/1693	Italy, Sicily	60000	7.5	
1268	Asia Minor, Silicia	60000		
06/20/1990	Western Iran	50000	7.4	Estimated 40,000 to 50,000 people killed, more than 60,000 injured, 400,000 or more homeless, and extensive damage and landslides in the Rasht-Qazvin-Zanjan area, Iran. Nearly all buildings were destroyed in the Rudbar-Manjil area. Substantial damage occurred as far away as Khalkhal and Now Shahr, and slight damage occurred at Tehran. Felt in most of northwestern Iran, including Arak, Bakhtaran, and Tabriz. Slight damage also occurred in southern Azerbaijan, USSR. Felt (VII) at Astra and Lenkoran; (VI) at Dzhibrail, Lerik, Mossony and Yardyshny; (III) at Baku, USSR. Complex event.
02/04/1783	Italy, Calabria	50000		

Note: Some sources list an earthquake that killed 300,000 people in Calcutta, India, on October 11, 1737. Recent studies indicate that these casualties were most likely due to a cyclone, not an earthquake.

(Source: "The 1737 Calcutta Earthquake and Cyclone Evaluated" by Roger Bilham, *BSSA*, vol. 84, no. 5, 1650–1657, October 1994)

Data compiled from several sources.

Appendix 4

Glossary

*Definitions supplied from various sources as indicated.

catastrophic geophysical event (CGE). Any one of numerous naturally occurring events that result in, or threaten to result in, a cataclysmic movement of matter and/or energy release, causing significant loss of human life along with an associated widespread destruction or alteration of the geology around the event location. (IEVPC)

Types of CGE include:

1. Volcanic eruptions
2. Tsunamis, regardless of cause
3. Earthquakes
4. Landslides
5. Floods
6. Droughts
7. Pathogens

8. Asteroid or cometary impacts

9. Material or radiation impacts on the Earth originating from the Sun or other bodies in the solar system (e.g., solar flare, coronal mass ejection).

10. Material or radiation impacts on the Earth caused by forces outside the solar system

earthquake. A trembling of the earth caused by a sudden release of energy stored in subsurface rock units. This release of energy usually occurs when the subsurface rock units break to form a fault or when movement on an existing fault occur. (Geology.com)

hotspot, geological. A volcanic center located within a lithospheric plate that is thought to be caused by a plume of hot mantle material rising from depth and located above a "hot spot" on the outer core. Examples of hotspots are the main island of Hawaii and the island of Reunion in the southwestern Indian Ocean. (WIKI)

grand minimum. A period of deep decline in solar activity as measured by sunspots that typically lasts two or more 11 year solar cycles such that the average of each cycle at peak reaches 50 or fewer sunspots. (Veritence)

magnitude (of an earthquake). The magnitude is a number that characterizes the relative size of an earthquake. Magnitude is based on measurement of the maximum motion recorded by a seismograph. Several scales have been defined, but the most commonly used are (1) local magnitude (ML), commonly referred to as "Richter magnitude", (2) surface-wave magnitude (Ms), (3) body-wave magnitude (Mb), and (4) moment magnitude (Mw). Scales 1–3 have limited range and applicability and do not satisfactorily measure the size of the largest earthquakes. The moment magnitude (Mw) scale, based on the concept of seismic moment, is uniformly applicable to all sizes of earthquakes but is more difficult to compute than the other types. All magnitude scales should yield approximately the same value for any given earthquake. (USGS)

Modified Mercalli Scale. The Modified Mercalli (MM) Intensity Scale was developed in 1931 by the American seismologists Harry Wood and Frank Neumann. This scale is composed of increasing levels of intensity that range from imperceptible shaking to catastrophic destruction, is designated by Roman numerals (ranging from *I* to *X*, where *X* represents extreme shaking).

It does not have a mathematical basis; instead it is an arbitrary ranking based on observed effects.

The Modified Mercalli Intensity value assigned to a specific site after an earthquake has a more meaningful measure of severity to the nonscientist than the magnitude because intensity refers to the effects actually experienced at that place.

The *lower* numbers of the intensity scale generally deal with the manner in which the earthquake is felt by people. The *higher* numbers of the scale are based on observed structural damage. Structural engineers usually contribute information for assigning intensity values of VIII or above. (USGS)

multicycle pause. A period of reduced solar activity or solar minimum within a cluster of 11-year solar cycles that is easily recognized for its much lower sunspot average at the peak of the cycle. This pause, essentially a special solar minimum and a rare form of the 11-year solar cycle, normally signifies the center of the cluster. (Veritence)

solar cycle. (a.) A period of 11 years, on average, during which solar activity measured by sunspots, varies from a low number (often near zero) to a peak (sometimes 150 or more sunspots) and then declines back to near zero as the next cycle begins. (b.) A general period of time during which the Sun's energy output varies from a high to a low point. There are numerous solar cycles of varying lengths including the best known of 11 years, followed by a 22 year, 55-60 year, 90-100 year, 206 year, up to the approx. 110,000 year ice age cycle. (Veritence)

Solar Index and Long term Sunspot Observations (SILSO). The web site and organization (Solar Influences Data Center) that compiles sunspot and other solar activity measurements for the Royal Observatory of Belgium, one of two internationally recognized sunspot number tracking organizations. NOAA is the other sunspot tracking organization. (Veritence)

solar minimum. (a.) A decline in solar activity as measured by sunspots during the normal 11-year solar cycle when the average smoothed 12-month average reaches its lowest number. (b.) A general period of declining solar activity among several 11-year solar cycles. (Veritence)

solar hibernation. A period of dramatically reduced energy output by the Sun that lasts for 20–30 years or longer. It is defined by sunspot counts during the normal 11-year solar cycle that average in the 50s at peak of the cycle compared with normal peak counts of 100–200. These hibernations come around roughly every 206 years. (Veritence)

These periods of reduced output of the Sun are also measured by lower levels of total solar irradiance (TSI), lesser solar wind, weak auroras, lower levels of radio signals from the Sun along with much slower surface movement on the Sun's surface. In addition, it is observed by a reduced solar magnetic field and lower Earth-Sun magnetic field. During these times because of the weak solar wind, cosmic rays from outside the solar system bombard the Earth's atmosphere which, according to some researchers causes more clouds, and thus a cooler surface and atmosphere for our planet.

sunspot. A region on the surface (photosphere) of the Sun that is temporarily cool and dark compared to surrounding areas. (NASA)

sunspot number. The sunspot number is calculated by first counting the number of sunspot groups and then the number of individual sunspots. The sunspot number is then given by the sum of the number of individual sunspots and ten times the number of groups. Since most sunspot groups have, on average, about ten spots, this formula for counting sunspots gives reliable numbers even when the observing conditions are less than ideal and small spots are hard to see. Monthly averages (updated monthly) of the sunspot numbers show that the number of sunspots visible on the sun waxes and wanes with an approximate 11-year cycle. (NASA)

Note: there are actually at least two "official" sunspot numbers reported. The International Sunspot Number as compiled by the Solar Influences Data Analysis Center in Belgium has been revised recently (V2.0, summer 2015) and should now more closely match the NOAA sunspot number. The NOAA sunspot number is compiled by the U.S. National Oceanic and Atmospheric Administration. The numbers tabulated are the monthly averages of the daily sunspot number with error estimates as posted at the WDC-SILSO, Royal Observatory of Belgium, Brussels.

tectonics. A branch of geology concerned with the structure of the crust of a planet (as earth) or moon and especially with the formation of folds and faults in it. (NASA) tsunami. A large gravity wave produced by a sudden displacement of a large volume of water. The displacement is usually caused by an earthquake, but it can be caused by submarine landslides, subaerial landslides that enter water, explosive volcanic eruptions, caldera collapses, iceberg calving, and asteroid impacts. These events suddenly depress or elevate a large volume of water, then gravity causes the energy of that displacement to propagate away from the source at a high rate of speed, often as fast as 500 miles per hour and often travelling across entire ocean basins. The waves have a very long wavelength of up to 100 miles, but their amplitude is typically so low that they can travel beneath ships without being noticed. Most tsunamis originate in the ocean, but they can be produced in lakes, bays, and rivers. When they enter shallow water, the energy of the wave begins to drag on the bottom and that slows the front of the wave while the back of the wave piles up behind it, reaching heights of up to 100 feet. (Geology.com)

Some people use the term *tidal wave* instead of *tsunami*, but that is incorrect because a tsunami has nothing to do with tides. The term "seismic sea wave" is correct if the tsunami is produced by an earthquake.

total solar irradiance (TSI). A measure of the Sun's energy over all wavelengths per unit area incident on the Earth's upper atmosphere. TSI has two ranges of measurement as a function of which satellites are being used. The most common measurement has been an average of 1366 watts/meter squared at the top of the Earth's atmosphere. A newer SORCE satellite (2003) uses an average of 1361 W/M squared. (Veritence)

United Nations Intergovernmental Panel on Climate Change (UN-IPCC). The primary body under the UN Environmental Development Program (UNDP) chartered to determine the effect of mankind's CO_2 on climate change. The UN-IPCC was formed in 1988 and has published its assessment of mankind's impact on climate change every few years since its first report in 1990. The last report was issued via several elements between 2013 and 2014. (Veritence)

volcano. A vent in Earth's surface through which molten rock and gases escape. The term also refers to deposits of ash and lava that accumulate around this vent. (Geology.com)

volcano explosive index (VEI). A method of comparing the severity of explosive volcanic eruptions using the volume of material ejected as a scale. The scale is logarithmic and begins at 0 for an eruption that produces less than 0.001 cubic kilometer of ejecta. Each step in the scale is a 10X increase in the volume of ejecta. About fifty eruptions have been rated at the highest VEI value of 8 on this scale, producing 1,000 cubic kilometers of ejected material. For comparison, the 1981 Mt. St. Helens eruption had a VEI of 4, Mt. Pinatubo was VEI 5, Krakatao was VEI 6, Crater Lake was a VEI 7, and Mt. Toba was VEI 8. Other supervolcanoes ejected much larger amounts than Toba, but all are in the VEI 8 class. (Geology.com)

Appendix 5

References by Chapter

Chapter 1: Where Are We Today?

1. Kishor S. Jaiswal, Mark D. Petersen, Ken Rukstales, and William S. Leith, "Earthquake Shaking Hazard Estimates and Exposure Changes in the Conterminous United States," *Earthquake Spectra* vol. 31, no. S1 (December 2015) pp. S201-S220.

2. USGS "Earthquake Outlook for the San Francisco Bay region 2014-2043," http://dx.doi.org/10.3133/fs20163020.

Chapter 2: The Sun-Earth Connection

3. P. Reimar et. al., "INTCAL04 Terrestrial Radiocarbon Age Calibration, 0–26 cal kyr BP," *Radiocarbon*, vol 46, no. 3 (2004), p 1029–1058.

4. http://www.drroyspencer.com/2016/07/uah-global-temperature-update-for-june-2016-0-34-deg-c/

Chapter 3: The New Madrid Seismic Zone (NMSZ)

5. https://www.census.gov/history/pdf/wright-hunt.pdf

6. https://en.wikipedia.org/wiki/2011_T%C5%8Dhoku_earthquake_and_tsunami

7. A. S. Elnashai et al., "Impact of New Madrid Seismic Zone Earthquakes on the Central USA, Volume I," MAE Center Report No. 09-03, October 2009.

8. https://en.wikipedia/wiki/huricans_Sandy_relief_bill.

Chapter 4: The West Coast: Widespread Damage

9. "Third Uniform Earthquake Rupture Forecast." http://www.WGCEP.org/UCERF3.

10. C. Nelson, C. H. Morey, A. E., Johnson, J. E., Patton, J .R., Karabanov, E. Gutiérrez-Pastor, J. Eriksson, A.T. Gràcia, E. Dunhill, G. Enkin, R.J. Dallimore, and T. Vallier, "Turbidite Event History—Methods and Implications for Holocene Paleoseismicity Of The Cascadia Subduction Zone: U.S. Geological Survey Professional Paper 1661–F," http://pubs.usgs.gov/pp/pp1661f/.

11. Alejandro Lazo, "Leaning Condo Tower Raises Alarm," *The Wall Street Journal*, September 23, 2016.

12. Associated Press, "Reaching High Into the Western Sky," *The Wall Street Journal*, September 6, 2016.

Chapter 6: South Carolina, Puerto Rico, and Hawaii

13. . Carl A. von Hake, Earthquake Information Bulletin, volume 8, no. 6 (abridged), November–December 1976.

14. . South Carolina Emergency Preparedness Division, "Comprehensive Seismic Risk and Vulnerability Study for the State of South Carolina," 2001.

15. . R. L. Dart, Pradeep Talwani, and Donald Stevenson, "Earthquakes in South Carolina and vicinity 1698–2009: U.S. Geological Survey Open-File Report 2010–1059," 2010.

16. . P. Talwani and W. Schaeffer, "Recurrence rates of large earthquakes in the South Carolina Coastal Plain based on paleoliquefaction data," Journal of Geophysical Research, vol. 106, no. B4, April 10, 2001, p. 6621–6642.

17. . http://www.decodedscience.org/puerto-rico-earthquake-m6-5-january-2014/41694.

18. . https://www.pnsn.org/blog/2016/01/07/slow-earthquake-trembles-beneath-vancouver-island.

Chapter 8: Preparing for the Great Upheaval

19. "Final Report of the Select Bipartisan Committee to Investigate the Preparation for and Response to Hurricane Katrina," February 15, 2006, page 246.

20. http://floridapolitics.com/archives/221639-john-mica-declares-fema-response-louisiana-not-acceptable.

21. Stephen Self, Jing-Xia Zhao, Rick E. Holasek, Ronnie C. Torres, and Alan J. King, "The Atmospheric Impact of the 1991 Mount Pinatubo Eruption," http://pubs.usgs.gov/pinatubo/self/.

22. Ted Koppel, *Lights Out: A Cyberattack, A Nation Unprepared, Surviving the Aftermath*, Crown Publishers, 2015, page 15–16.

23. "ShakeAlert: Earthquake Early Warning," http://www.shakealert.org.

24. https://en.wikipedia.org/wiki/Northeast_blackout_of_2003.

25. http://tobyhemenway.com/419-fear-and-the-three-day-food-supply-3/.

26. http://www.businessinsider.com/how-many-days-can-you-survive-without-water-2014-5.

27. David Crook and Louis A. Santiago, "Hoe Big Is the Hunting Economy," *The Wall Street Journal*, October 13, 2106.

28. http://www.fns.usda.gov/sites/default/files/pd/34SNAPmonthly.pdf

29. https://www.gobankingrates.com/savings-account/62-percent-americans-under-1000-savings-survey-finds/

30. http://www.bls.gov/news.release/empsit.nr0.htm

31. https://news.usc.edu/86362/fukushima-disaster-was-preventable-new-study-finds/.

32. http://www.climatedepot.com/2015/11/04/no-global-warming-at-all-for-18-years-9-months-a-new-record-the-pause-lengthens-again-just-in-time-for-un-summit-in-paris/

33. "Food and Water in an Emergency," FEMA Publication 477 A5055, August 2004, https://www.fema.gov/pdf/library/f%26web.pdf.

CPSIA information can be obtained
at www.ICGtesting.com
Printed in the USA
LVHW020217031118
595826LV00036B/575/P

9 781490 779034